ROUTLEDGE LIBRARY EDITIONS: POLITICAL GEOGRAPHY

Volume 8

T0251163

GEOPOLITICS

GEOPOLITICS

PATRICK O'SULLIVAN

Routledge
Taylor & Francis Group

LONDON AND NEW YORK

First published in 1986

This edition first published in 2015
by Routledge
2 Park Square, Milton Park, Abingdon, Oxon, OX14 4RN

and by Routledge
711 Third Avenue, New York, NY 10017

Routledge is an imprint of the Taylor & Francis Group, an informa business

British Library Cataloguing in Publication Data
A catalogue record for this book is available from the British Library

ISBN: 978-1-138-80830-0 (Set)
eISBN: 978-1-315-74725-5 (Set)
ISBN: 978-1-138-81029-7 (Volume 8)
eISBN: 978-1-315-74961-7 (Volume 8)
Pb ISBN: 978-1-138-81054-9 (Volume 8)

Publisher's Note
The publisher has gone to great lengths to ensure the quality of this reprint but points out that some imperfections in the original copies may be apparent.

Disclaimer
The publisher has made every effort to trace copyright holders and would welcome correspondence from those they have been unable to trace.

Printed and bound by CPI Group (UK) Ltd, Croydon, CR0 4YY

Geopolitics

Patrick O'Sullivan

CROOM HELM
London & Sydney

© 1986 Patrick O'Sullivan
Croom Helm Ltd, Provident House, Burrell Row,
Beckenham, Kent BR3 1AT

Croom Helm Ltd, Suite 4, 6th Floor,
64–76 Kippax Street, Surry Hills, NSW 2010, Australia

British Library Cataloguing in Publication Data

O'Sullivan, Patrick
  Geopolitics.
  1. Geography, Political
  I. Title
  320. 1'2   JC319

   ISBN 0–7099–1934–4

Filmset by Mayhew Typesetting, Bristol, England
Printed and bound in Great Britain by Mackays of Chatham Ltd, Kent

CONTENTS

LIST OF FIGURES

1   CONFLICT, TERRITORY AND DISTANCE

Geography and Politics

The premiss of this book is that geography matters in the relations bet-ween states, be they friendly or hostile. The strength of sympathetic and influential ties between the governments of nations is as much a matter of geographic distance as of political and cultural distance. Certainly in tracing both the causes and courses of wars, distance and space count for a lot at the grand strategic level. More locally in time and space, the details of geography provide the immediate setting for deadly con-flict and strongly affect the outcome of war.

Whether we like it or not, the crux of international politics is war. We punctuate history with the violent contests of powers. The immediate circumstance of these outbreaks of killing and destruction is most usually contested turf. Political authority is conceived of as extending over some portion of the earth's surface and the competition of those with power is over whose writ runs where. The domains of significance obviously extend beyond the limits of the nation state and the most serious con-flicts today, from the viewpoint of man's survival, arise in the ill-defined setting of the spheres of influence of the superpowers.

Even if we account importance in the more mundane terms of the use of our resources, we spend a lot of time, effort and material in fighting each other or getting ready to do so. The number of people whose lives are dedicated to war, to making weapons and the exertions and ingenuity put to these ends, uses up a large share of the world's wealth. In our political dealings with each other we frequently suspend the moral rules we ordinarily live by and are willing to kill and maim other people and destroy their property, or to threaten these things, in order to enable our leaders to get their way. Violence has been a ready resort and the preparation for war is a major feature of most civilised societies. Old men prepare and induce young men to kill the enemy. In the past the enemy was usually other young men but now it includes women, children, the old and the earth itself. The main reason for going to war has been to command more territory, and this remains the chief source of conflict. The greatest danger to our lives and well-being does not arise from the abstract clash of ideologies, but from the collision of the forces of the great powers or their clients over some piece of land. Whether the

1

self-justifying rhetoric which accompanies such clashes is 'revolutionary' or 'democratic', the occasion of conflict is the control of territory. The fundamental dimension of conflict between nations is geographical. Geopolitics is the study of the geography of relations between wielders of power, be they rulers of nations or of transnational bodies. The limiting state of these relations is war and, no matter how we may abhor its wastefulness, this has been the main means of resolving the competitive process. War provides the negative image of affairs between peoples, and thus delineates the extent of peaceful co-operation or competition in the world.

There are now over 4 million people from 45 of the world's 164 nations involved in combat, with up to 5 million people killed in fighting over the last three years. There is a concentration of turmoil at the pivot of the continents where Africa and Eurasia meet and a wider scatter beyond in Africa, South Asia and Latin America. The mainsprings of these outbursts of bloodletting and fire are mostly local. They are sparked by the frictions of social, economic and political change and personal ambition and are not in essence elements in any worldwide plan for conquest and domination. At times these local conflicts have been exploited by the government of one powerful state to embarrass that of another power. Or politicians and officials have sought to cover up their own lack of foresight, their ineptitude or inability to control affairs in their client states, by claiming that the troubles were mischievously generated by their protagonists as part of their campaign of conquest. Spokesmen for the USSR have claimed that the American government is behind the Afghan mujahedin, while Jeane Kirkpatrick put the El Salvador guerrilla war in somewhat confused global context when she announced that 'an eastern offensive on our southern borders' was not tolerable to 'the west'. Some holders of power and office may actually believe in the simple theories of global politics in which they couch their foreign policy pronouncements. Whether these anathemas hurled at the opposition arise from a heartfelt belief or are merely a face-saving device, they contort the big powers' attitudes to others' conflicts. These are treated not on the basis of judgement of local rights and wrongs, but as an outlet for ambition and competitive strutting.

Mental Maps

In reacting to tension and strife around the world, the foreign policy establishments of the great powers do seem to have generalised images of political geography in their minds. These images are occasionally articulated as powerful, allegorical maps. The lines of their divisions of the globe clearly transcend the frontiers of nation states. Maps of political geography delineating the world's 164 countries are obviously missing something. For the politicians and functionaries of the great powers there are, at least, mental maps of 'our' and 'their' territory which supersede national boundaries, with an anyman's land lying between them. It is evident that the lines of demarcation on these mental maps do not coincide as between the two main camps, and the overlap of claims brings the dangerous might of the continent-spanning powers too close to each other for everyone's comfort.

The Economist 26 December, 1981 offered an analysis of 'the East–West struggle' illustrated by maps depicting two levels of either pro-Soviet or pro-American leaning and a fifth state of non-alignment or contested affinity. This article was a contribution to the debate about the proper preparations for war against the USSR by the USA and some Western European nations. In this debate the respective government objectives of East and West and their expectations about their rival's global ambitions are voiced in a vague fashion.

More militant Western leaders of opinion condemn the Soviets for expansive, territorial aggression while crediting the US government with a defensive, altruistic attitude. The defence of 'freedom' and a foggy, enveloping notion of US 'national interest' are sufficient, however, to warrant intrusion, direct or indirect, into the provinces of the opposition or the undecided lands between. The image of spheres of influence around the rival poles of the USSR and USA continues to frame the rhetoric of this competition. Since 1950 economic and political reality has departed further and further from the picture of bipolar contention, even though nuclear confrontation has remained essentially duopolistic.

It is, of course, feasible to draw a map, such as Figure 1.1, of the domains over which the USA and USSR could be said to exert political leadership right now, reflecting the current of informed opinion and news, putting nations into one camp or the other, or leaving them unaligned. Such a bald statement of the two hegemonies would invite objection, exceptions and denials. Whether North Korea is inclined to Russia or China right now; whether Qadaffi is a creature of or embarrassment to the USSR; whether France's premier is feeling Atlantic or European today; whether Soviet service will pay off in Iran or whether President Machel's submission to South Africa signals a switch of camps for Mozambique — all are considerations open to dispute and will change the colours of the map from day to day. Such a map is a snapshot of

Figure 1.1: Spheres of Influence

US Sphere
Non-aligned
Soviet Sphere

an oscillating, evolutionary process subject occasionally to catastrophic change. Thus, it would have no lasting relevance. Yet it does seem from the words of statesmen and the imagery of their language that some mental construction of similar content and intent provides an ordering of their world and a matrix for the most important decisions made on the fate of mankind. Lines are drawn, spheres defined and encroachments contained on a projection of the globe, expressing politicians' fears and designs. The scoreboard for the contest of their ambitions is a map.

The three-toned simplicity of a map of the hegemonies is of little value other than to feed paranoia with a crude image of the political instant. To express the nuances and ambiguities of relations among nations with accuracy it would take many more than even the five shades of *The Economist's* maps. The military dominance of the USA and USSR is beyond dispute, but the strength of their influence has been pushed and pulled back and forth in China, Southeast Asia and Latin America since 1950, while in the Middle East and Africa the two superpowers have probed and intertwined in tentative compensation for the European stand-off.

Geopolitics

What any mapping of the realms of power does suggest, however, is that the force of international influence does still reflect the tyranny of distance, despite the protestations of Wohlstetter and Bunge. The more removed a nation from either great power, the better the prospects of remaining neutral. A variety of theories can be invoked to explain why the magnitude of projected power should diminish with distance. The main purpose of this book is to explore this attenuation theoretically and empirically and its implications for international affairs. To signal this concern with matters of distance and place, I have used the title *Geopolitics*. This implies the use of geographical good sense in understanding or governing the relations between groups of people. If politics is the art of government, then the 'geo' prefix implies the application of geographical knowledge to this end. The term is usually applied to foreign policy, but clearly some domestic matters could benefit from the exercise of geographical nous.

The main service a geographer can provide in foreign affairs is to emphasise the defence which geography offers against the ambitions of rivals and to instil as much objective, geographical reality into the dealings between nations as possible. Distance, difficult terrain and the sea still provide obstinate barriers to fighting men and their machines. The space and surfaces between the urban cores and resource bases of the USA, USSR, Western Europe and China are big and awkward enough to afford them confidence that one will not attack the other directly with a view to conquest. The main dangers to peace, however, are seen to lie in the vulnerability of their clients and neighbours to indirect disruption. In dealing with these contingencies the strategies, ploys, doctrines and myths of rulers, functionaries and popular expressions of opinion frequently display departures between geographical reality and the mind maps which chart power and danger in foreign affairs. Fear and hate are fed by misperceptions and misapprehensions of geography.

The two main themes of this book, then, are the gaps between geographical myth and reality, and the persistent potency of distance in checking vaulting ambition.

Motives

To deal with these matters adequately we must first consider the motives of those who decide on and operate the relations between nations and their views of the world. This is the subject of Chapter 2. It has proved impossible to codify the motives of makers of foreign policy satisfactorily in order to produce a theory which will stand the usual tests of validity. Time after time events deny the assumption of simple utilitarian or altruistic objectives or, indeed, any singular rationale. There does, however, seem to be a persistent lust in some people to control the lives of others. Beneath the veneer of ethical and ideological rhetoric which has been used to justify actions, history records the results of individuals and groups giving rein to what appears to be an innate and widespread domineering impulse. Luther called this the *animus dominandi*, the lordly spirit. If it is not curbed by humility and self-discipline or checked by others, it can result in dominions that encompass great tracts of the earth. To generalise on the geographical arrangements of political life, it seems we must accept this aggressive spirit, which is widely accepted by the players in the international game themselves, as the fundamental driving force. From this viewpoint any failure to exploit a competitive advantage or another's territorial withdrawal is interpreted as a ploy to regroup strength for future expansion or as a sign of weakness.

Perceptions

History also suggests that our rulers are as prone as the rest of us to misconstrue the nature of the world and the intentions of others. Many trains of events can only be traced to blundering stupidity and ignorance. The uncertainty which has plagued the dealings between the USA and USSR since the 1940s has arisen not only from the lack of correspondence between the official pronouncements and action of both sides, but also from the departures between the underlying objectives and actions — not only from duplicity, but also from errors of perception. Chapter 3 explores the variety of perceptions of this world.

To explore the perceptions which guide policy we begin with the variety of traditional and classical views of the world and the deep persistence of a sense of territoriality. The feeling of nationhood contrasts with a sense of belonging to broader groups such as Christendom or Islam. The visions and problems of imperialism and global hegemonies lead us into the geopolitics of Mahan, Mackinder and Haushofer.

To trace the origins of the views of the world which underlie today's marking-out of spheres of influence, it is necessary to delve into the geopolitical context of Marxism; Lenin's changing geographic perceptions; the synthesis of traditional Russian and ideological attitudes; Maoist rhetoric and age-honoured Chinese cosmology. As opposed to these visions, the overt imperialism of Britain and the Pax Britannica provided both a model and a target for the confusion of motives involved in the US rise to hegemony. The Monroe Doctrine, 'democratic' ideology and economic imperialism are some of the threads in an evolving pattern of American thought, speech and behaviour in foreign affairs. The 'spheres', 'dominoes', 'arcs' and 'chains' which crop up as images of political geography and the conflict with communism for influence over the world, have to be understood against this background. The ambivalence of the makers of policy over the Pax Americana and imperial responsibility is perhaps the most fascinating theme in international relations. While the USSR and China have reverted to traditional and comparatively static roles in recent years, there is a powerful body of opinion in the USA which advocates overturning the Russian polity. Foreign policy for these people is spoken of as a crusade to destroy communism and make the world safe for democracy. Containment, then, is a temporary expedient. In previous administrations the announced position was not so clearcut, teetering on the sharp edge formed by power and responsibility. This uncertainty as to the proper position to take has caused bewilderment and unease over the drawn-out and sometimes violent contest for *de facto* spheres of influence between the USA and USSR. Within the traditional alignments of US politics there is a fundamental division of geographical attention between the Pacific and Asia and the Atlantic and Europe. This dilemma reflects a tension between imperial and more co-operative tendencies. One mood of American politics was inclined to accelerate the retreat of European colonialism after 1945. Another humour tried to fill the political vacuum created by this withdrawal in an effort to contain communism. In the case of Indochina this culminated in a retreat of confused pain. Americans were shocked to discover a lack of correspondence between their perceptions of the state of the world and their role and the reality.

Theories

Chapter 4 looks at some theories of international competition and conflict. Making a theory involves describing phenomena and their interplay to understand how they behave, in the hope of predicting this behaviour. Thus, theory requires the representation of the things in question and the naming of the driving force which powers their movements. There are obviously different views on the motive force underlying global politics and a variety of ways of representing international relations with a view to explaining them. Of late there have been efforts to reconcile the classical realist view that the primary objective of foreign policy is security and power with the more recent welfare school of thought. The apparatus of economics has been adapted to international politics, defining the state as an organisation for the provision of protection and welfare in return for revenue. The people who constitute the state are seen as making corporate, hedonically rational decisions in their competition for resources and in their strategies. Such theorising could be made spatial by the introduction of distance and direction into its formulations. International politics could be viewed as an exercise in net benefit maximisation with determinable territorial equilibria, optimal scales of operation and desirable levels of commercial intercourse and military involvement. The dynamics of the international system could be treated as a matter of adjustments to technical change in communications, production and military capabilities. If location were incorporated as a variable in this framework, the theory could be rendered geographical by applying different costs of friction over land and sea, plain and mountain, grassland and forest. Terrain and channels of movement might be made explicit factors in the theory. But no matter how well and detailed the condition of the scenery could be laid out in theorising of this kind, its validity stands or falls on the grounds of the motives it ascribes to politicians. The evidence is that no generally acceptable conclusion is likely to be reached on this.

If we are willing to make a crassly simple assumption about the objectives of foreign policy, then it may be possible to produce a crude first approximation to geopolitical behaviour, in either deterministic or probabilistic fashion. The essence of the state is territoriality and the focus of international competition is control of territory. To simplify matters further to get a starting-point, geography can be reduced to a line. Competition for territory among a row of states along a line offers a first step. From this elementary beginning more elaborate representations might be built up.

In Chapter 4 we focus on two types of theory about the relations between nations. The first type concentrates on the decision-maker and sacrifices the geographical context to elaborate the calculus of statesmanship. This category includes the economic and game theory models of international affairs. If we wish to introduce the reality of space into our theorising about the politics of nations, we are forced into a very spare representation of political motives and choices in the second category, spatial theories.

Distance Friction

Whatever the purposes of exerting might beyond a country's boundaries, the potency of this projected power does appear to be reduced by distance from its generating source. Chapter 5 examines the reasons why this should be. In trying to build a geographically specific picture of the disposal of power and its changing balance, distance is the fundamental dimension involved. The diversity of mankind was fostered by the frictional effect of spaces and surfaces which separated gene pools and cultures. Since the European conquest of the globe in the sixteenth century, there has been an erosion of the impediment of distance and a convergence of culture realms drawing us all into a single interacting ecumene. The separating power of space has been diminished. Some would go so far as to say that with electronic communication and command of flight, distance has been abolished as a significant factor in world politics. Certainly it is still with us in our daily lives, separating our beings and doings. Despite the frantic efforts of shuttle diplomacy, even in the affairs of states distance still inhibits interplay — for better or worse. Indeed, it might be that what can be regarded as tyrannical as far as reducing trade and social interaction is concerned, may serve as a blessing in keeping political ambitions apart. In military terms certainly, distance is still the best defence.

The most compelling picture of the influence of distance on power is conjured up by Boulding's 'loss of strength gradient'. This notion expresses the idea that power is greatest at home and is lower the longer the distance from the home base. The farther force is extended, the weaker it becomes. Besides the frictional effect of the cost of overcoming distance, the energy expended in controlling a widening expanse of territory and the lack of familiarity which goes with distance from the homeland may well have their effect on national morale and willingness to extend influence.

Boulding's explanation for the weakening of might with distance was that the cost of distance friction in transporting force and sending communications eroded the magnitude of power. The development of naval power, then radio, air power, rocketry and satellites has diminished this gradient to the point where there are those who would dismiss distance as a factor in the balance of power. The exertion involved in putting together a Rapid Deployment Force for the US to operate in the Middle East suggests, however, that distance still requires effort. Some of the trade-offs between force and distance and the difficulties of 7,000 mile supply lines were revealed in the Falklands War. In as much as trade between nations does display a tendency to attenuate with distance, the potential for the exertion of economic clout must also be dampened in a similar fashion.

Besides the frictional effect it is possible to think of other reasons why the potency of power should diminish with the radius over which it is exerted. If we think in terms of either an empire or an hegemonic power, the essence of its domination is that it exercises power over territory, that its rulers can impose their will by their superiority of force and prestige. In the final analysis the military capacity to control territory is the measure of power. Hegemony does not imply the complete political integration of a sphere of influence, but the capacity to exclude a rival and dictate what writ runs in a territory does require that the force to control it be available when it comes to the crunch. If the core state of an empire or an hegemonic state has limited capacity to deploy men and material, then as the margins of its territorial ambitions expand, so the same force must be spread more thinly and lose its potency. If it were expanding symmetrically on an even plain, then the potency of power would attenuate at a rate of 2π for every extension of the imperial radius. The same force would be spread over wider and wider circles and its density on the ground and strength would be diminished accordingly. To put it another way, if the cost of control per unit area is the same everywhere, as the circumference of an empire expands so the total cost of controlling it will increase as the square of the radius. In order to maintain a certain level of presence everywhere, as the circle of empire expands incrementally so the military establishment must increase exponentially. Whether or not there is a frictional effect, there will certainly be a thinning effect on military force with more far-flung ambitions or obligations. The same effect would apply to propaganda and subversion or economic aid. The Kremlin seemed to have been swayed by such considerations in discouraging the extension of the Cuban revolution to the American mainland, judging that it could not afford another Cuba.

In addition to the material friction of distance there does seem to be a gradient of boldness which operates for individuals and groups. In his description of the territorial rites of the stickleback, Konrad Lorenz points up a loss of courage far from home as the cause of a falling off of power with distance. The boundary between the domains of two male sticklbacks is set by a series of attacks and counterattacks with the resolve of each waning as he gets further from the nest he built. The battle wages to and fro with shallower and shallower attacks until a balance of their courage results in a stand-off along some intermediate boundary. Such a loss of courage or of moral certainty at a distance seems to have been at work on the performance of the USA in Southeast Asia and the USSR in the Caribbean.

Any of these effects or a combination of them provide a mechanism with which to theorise about competition among poles of power for dominion over the globe. The loss of strength gradient is employed in Chapter 5 as a cost function to produce results similar to those of central place and rent theory in locational economics.

Accepting the fundamental limits to conjecture arising from simple suppositions about the intent of statesmen, these abstractions may provide a fresh perspective on a subject usually cluttered with personalities and minutiae.

One thing which this kind of enterprise provides is better-defined language with which to discuss competition among the powers for dominion over the globe. Changing capabilites, expansion and contraction can be viewed in terms of force fields which diminish in intensity with distance from the projecting foci of power. The political surface then consists of poles of power with their diminishing force fields draped over them like bell tents, intersecting with each other in accord with the heights of the poles and their locations. The power pole whose power field is higher at any location dominates that part of the globe. This gives a more precise meaning to the notion of a sphere of influence. We can think in terms of a mapping of dominated territory with the great extent of the hegemonies' spheres of influence separated perhaps by intervening independents and possibly interrupted by isolated, autonomous states. Such a force field image of the dealings between nations may not be too fanciful. It seems to correspond with the conception of some practitioners of the art at least. In 1944 Churchill proposed to Stalin that postwar spheres of influence in southeast Europe be defined by a percentage plan. The UK was to have 90 per cent influence in Greece and the USSR 10 per cent. In Hungary each was to have 50 per cent influence. In Romania the Soviets would have 90 per cent influence and the UK 10 per cent. There is no record of Stalin's reaction. The proposal does, however, bring to mind a picture of intersecting gradients of power. Over the long haul with the increase of its power in this century, the limit of the USA's declared sphere of influence has been explicitly extended. The Truman Doctrine pushed the outer boundary from the Monroe definition across the Atlantic and the Pacific. In the last few years Carter declared the US interest in the shores of the Persian Gulf and the Reagan Doctrine, embedded in the 1985 State of the Union address, proclaimed support for anti-communist insurrection wherever it arises.

Even though it lends itself to theoretical manipulation and corresponds in a fashion to the perceptions of statesmen or official declarations of territorial rights, to think of the political world as consisting of continuous fields is a distortion of power. The image of power disposed as a continuous field draws notice to the idea of domination and the places from which superior power emanates. A little reflection on our interactions with one another suggests that they are more like channelled flows than waves. Our human numbers and the links between us are finite. The means of exchange and communication between people are limited. Although television and radio broadcast events instantaneously around the world, when we respond to signals we do it through a countable number of links. The space in which mankind interacts operates more nearly like a finite graph than a force field, especially at the international level. There is a network of a limited number of connections joining some nodes of power to others. Interaction occurs along the link of this network, taking both direct and indirect paths. The magnitude of power transmitted is the resultant, not only of the prestige of the sender, but also of the receptiveness of the receiver and the friction along the path between them. Nodes in the net coincide not only with national governments, but can be regional or transnational sources of power.

Measurement

Having postulated some relationships and possible geopolitical structures, we turn in Chapter 6 to the problems of measuring the ties and antipathies between nations and the disposition of power.

To measure power and the flow of influence between countries is a tall order. Power is exerted through armed might, diplomatic channels, the manipulation of economic relationships and the propagation of ideas. A complete mapping of fields of influence would involve measuring military, diplomatic, economic and cultural relationships between groups of people in different parts of the world. Consistent data are only available for transactions between states. It is not possible to break down the places where choices and decisions are made finely enough to quantify the intranational or supranational dealings in power. Although economic deals are made between large numbers of firms or agents, they are statistically recorded on a national basis. There are in addition two components of national power which are in principle indivisible and identified with government; these are armed force and diplomacy.

Of the variety of ways in which one nation can intrude on the other, including military intervention, subversion, treaties, cultural activities, propaganda and manipulation of the UN, probably the volume of trade and commerce is the most concrete and unambiguous expression of mutual interests. Although some people would decry the benefits of some types of trade in terms of unwanted dependency, there seems little doubt that a flow of goods balanced by a flow of payments is an immediate revelation of values by the population at large. The current pattern of international exchange, although it represents neither a political nor an economic equilibrium, provides the least equivocal map of shared interests among nations, least distorted by duplicity, fawning or show.

Trade Ties

In Chapter 7 International Monetary Fund figures for visible trade provide a basis for tracing fields of potential influence and links of mutual interest between nations. These data leave us in no doubt of the commanding position enjoyed by the US economy in the Americas,

Atlantic Europe, Southern Africa, the Middle East, India, Japan, the Philippines and Australia. By contrast, the Soviet economy dominates only Eastern Europe, Afghanistan and Mongolia, with Cuba and Ethiopia the only detached subordinates. The territory between, where neither the USA nor the USSR dominates, is broad, covering much of Europe, Africa and the southern tier of Asia.

These numbers describing the flow of goods between countries do make some patterns of economic and social change quite apparent. It is obvious that our political myths and metaphors need reappraisal in the light of clearly visible demographic and economic trends. Between the American and Russian core areas, economic vibrancy in the 1980s is focused on two major trading regions. One, centred on Western Europe, draws together Eastern Europe, Africa and Southwest Asia. The other, more explosive growth has Japan as its primary focus and links the margins of the Pacific, including the bustling economies of Taiwan, South Korea, Singapore and Hongkong. The USA plays a part in both of these, but does not overshadow their endogenous vitality. The conflict of interests between these two involvements and home-grown interests have and will provide a source of tension in US politics. The friction of distance in economic affairs does appear to operate so as to promote several regional groupings of countries at the global level. This reality does erode the image of the world as divided into two warring camps, an image which grew up in the 1950s and continues to colour the rhetoric and judgement of some people whose global consciousness was formed at that time.

War

If Chapter 7 establishes the positive pattern of common economic interest among nations, the final chapter turns to the shadows of distrust and conflict. The conflicting interests of governments have in the last resort been settled by war. This has been the chief means of conflict-resolution. The ability of an increasing number of military forces to spark nuclear fire makes this resolution of confrontations an appalling prospect for us all. No one can hope to win these in any meaningful sense of that word. The places where battles are under way now and the circumstances which started them are surveyed and the prospects of predicting future outbreaks are examined. What can be done to avoid war, apart from wishful thinking, is not apparent. But it does seem that the emergence of close-knit, prosperous, regional communities of interest which interlock to integrate the whole population into a co-operative ecumene, is a first step towards peace. This creates something for everyone to lose. As yet, the only mutually acceptable definition of common interest does involve material well-being and affluence generates pacific, middle-class dominance of politics. Although peoples have gone to war in denial of their own best economic interests, and although economic sanctions have not proved effective weapons to achieve political ends, nevertheless the links of trade and commerce are the fundamental minimum requirements of any international community of interest. We can only hope that the quickening of the international economy improves the prospects of reconciliation between the more extreme antipathies in the world.

Readings

There are two recent compendia of scholarly work on world politics and conflict:

Beer, F.A. *Peace against War: The Ecology of International Violence* (W.H. Freeman, San Francisco, 1981)
Russett, B. and H. Starr *World Politics: The Menu for Choice* (W.H. Freeman, San Francisco, 1981)

Much of the writing on international affairs, geopolitics and technology is put succinctly in its widest setting by:

Deudney, D. *Whole Earth Security: A Geopolitics of Peace* (Worldwatch Paper 55, Washington DC, 1983)

Distance was dismissed as a significant factor in international politics by:

Bunge, W. 'The Geography of Human Survival', *Annals, Association of American Geographers*, vol. 63, no. 3 (1973), pp. 275–95
Wohlstetter. A. 'Illusions of Distance', *Foreign Affairs*, vol. 46, no. 2 (1968), pp. 242–55

The gradients of power and courage come from:

Boulding, D. *Conflict and Defense* (Harper & Row, New York, 1963)
Lorenz, K. *King Solomon's Ring* (Signet, New York, 1972)

2 MOTIVES AND DECISIONS

Animus Dominandi

History is a striking record of the persistent desire of some people to lord it over others. Social arrangements and political institutions evolved to give authority to a few over the many. The events with which we mark the passing of time are mostly violent competitions for territory and authority. There does seem to be a persistent lust in some people, individually or in a group, to control the lives of others. This motive has a fair claim to be the driving force of history.

The style in which this quest for dominion is pursued varies according to the persons and the circumstances involved. Commentators on foreign policy have pointed to a strand of idealism, on the one hand, and of hard-headed calculation, on the other, running through the doings and sayings of individual leaders, never mind within administrations or within political elites. Since the ideals involved have the power of moral and emotional certainty behind them, we must credit idealism as the primary motive force. Pragmatic hard-headedness is a second-best position forced upon leaders by the constraints of circumstances and geography. Those who make pragmatism a principle will usually put it below some higher mission for their nation in the last resort.

The ideals of politics invariably involve convincing the rest of the world to behave properly. This is done by bringing other people under the will of the idealist, one way or another. In some cases, domination is seen as being good for the dominated. It increases their moral, and possibly material, well-being. To bring communism, civilisation, religion or the freedom of the market to others with a war of liberation, revolution, *jihad,* crusade or counterinsurgency operation, improves their lot. In other instances the best for all of mankind is evoked to warrant the subjugation or eradication of the dominated. It is an inevitable result of a natural process that some people be enslaved or killed to better the world and provide *Lebensraum,* or to fulfil the Manifest Destiny of a nation, or to satisfy the imperative of the historical dialectic.

Whether or not this is the fundamental force, opponents in international affairs will accuse each other of being driven by this spirit of conquest and at least posture as if they were themselves. It is very difficult to separate rhetoric and poses from the real basis of action and true feelings

16

at this level. Those who lead nations are, however, prone to delusions of greatness such that rhetoric and reality may become as one to them.. The greatest delusion does not concern your enemies' motives, however, but your own. The lust for power is justified when it assumes the cloak of righteousness. Beneath this mantle, the urge is to make the world over in your own imagined nature. This basic drive may be tempered and constrained by a calculating appreciation of the odds. Deference to reality, however, is merely a rational brake on consuming ambition.

US Foreign Policy

The avowed position of the US administration at the time of writing is that the leadership of the USSR consists of evil men who desire to rule the world. This is their nature and they cannot be diverted from this course permanently by any means other than destruction. Their spirit of conquest can be overruled only by wiping them out or keeping them in fear of death. The Soviet elite can be stopped from their thrusting only by the threat of overwhelming superior nuclear firepower. This has all the appearance of a firmly held belief.

The official announcements of the Soviet nucleus of power use similar rhetoric directed at the US military–industrial complex, when they find it convenient. It is even more difficult to discern the gap between belief and talk in the USSR, of course. Whether the motives it ascribes to the USA and the simplifications of propaganda are items of faith among the ruling elite, is not clear at all.

Trying to weigh the motives which have moved US foreign policy, as viewed from the sidelines, there do appear to be two seemingly contrary themes running through changes in circumstances and people. These are pragmatism and idealism.

The idealistic trait springs from Calvinist roots and sees the USA as a saviour nation, destined to bring the world to a right pattern of life. Any other nation could be forgiven for seeing a threatening aggressiveness in this. The American pragmatists pursued a narrower objective in terms of some definition of 'national interest'. This involved maintaining some balance of power in the rest of the world by undermining too great a concentration of might in others' hands. Even the founders of this posture, however — Washington, Jefferson, Hamilton and Adams — saw their nation as having a special mission. The purpose of balancing power was to protect the American experiment from interference from abroad so that it could eventually redeem the world, by example. Quite evidently, the more fundamental drive was to change the world, not merely to survive. The two threads then are not independent and contradictory, but only express different levels of resolution in viewing the world. The idealistic is a long-run, global viewpoint, while the pragmatic focuses on matters limited in time and space, but is always governed by the long-term goal.

In the mix of idealism and temporising shown by most leaders, the missionary motive may seem to be subordinated to power politics. But the spiritual nature of the ideal gives this moral superiority. When Woodrow Wilson brought the USA into the European war in 1917, he did not justify his action in terms of stopping the German concentration of power; instead, he spoke of the USA as the only idealistic nation, appointed by God to save the world from its bad old ways. In the last two-thirds of a century the USA has achieved productive and technical supremacy and has established a presence around the world. The accelerating intervention in foreign affairs of the USA in the 1940s was excused by Roosevelt on grounds of power politics, but it placed the USA's leaders in a position to pursue the ideological mission, with direct and indirect power over much of the globe. The main expression of the American ideal, however, came to take the negative form of hatred for communism. Pragmatism led to the support of regimes quite out of keeping with the ideal, in order to achieve the greater, long-term good which required the smashing of communism and the Soviet power apparatus. With Ronald Reagan we have a committed adherent to this inversion of idealism in command of the most powerful nation in the world.

Soviet Foreign Policy

The dogma of the USSR is quite explicit about the violent pursuit of revolutionary change. This more recent gloss on expansionist ambitions had its antecedents in Great Russian imperialism and messianism and pan-Slavism. This has always been tempered among the Russian elite by a fear and loathing of the non-Russian world and of Western Europe in particular, which finds expression in a strong isolationist tendency. Although dogma requires genuflection to revolution and the destruction of capitalist imperialism and drew the Soviets into far-flung foreign involvement in the 1950s and 1960s, a realistic and practical bent and deeper historical perspective among its present leadership have encouraged caution.

Both ideals and the material interests of nations have been cited to justify intervention in other countries' business. Trying to save the world obviously requires getting other people to be as you would have them, which involves imposing your will by force or influence. The ideological basis for policy then reduces to a desire to dominate other people and, thus, territory. The pragmatic inclination tempers this with calculation. The tendency to surge outward aggressively is held in check by considerations of domestic politics and the geography of opportunity and opposition. To quote Richard Nixon ('Real Peace: A Strategy for the West', *New York Times*, 2 October 1983):

> We will meet the challenge of real peace only by keeping in mind two fundamental truths. First, conflict is a natural state of affairs in the world. Second, nations only react to aggression when they believe they will profit from it. Conversely, they will shrink from aggression if it appears in the long run it will cost them more than it benefits them.

The Seat of Decision

The picture of aggressive, calculating rationalism applied to personified territory which Nixon presents does raise a number of questions about the decision-making process in international affairs. The aggression and recklessness of statesmen and the degree to which they employ consistent arithmetic in making choices are matters for historians, philosophers and psychologists. What geographers can usefully address themselves to in this realm is the identity of authority employed and its relation to territory.

The use of 'the nation' as the actor in statements about international relations involves territorial personification. There are three steps of abstraction involved in this. First, the nationalist myth which compounds a collection of people with the territory in which they live is implicit in this usage. Since the American and French Revolutions, the nation as a piece of territory under a particular sovereign, containing some number of people, came to be identified as 'the people'. The people were no longer spoken of as being in the nation but came to be the nation itself. The next step in this abstraction is the ascription of a singular will and set of desires to this population. By this means they become the collectivity 'the people' by dint of their residence on the same piece of land. Finally, the person or persons who actually wield the power of the state and choose for everyone else, the government, is identified with 'the nation', thereby gaining the authority of 'the will of the people' in the territory they rule.

It is evident that what at first appears to be a simple verbal convenience, honoured by its long-established use by historians and journalists, does involve fallacies of composition and identity. The effect is to inflate the authority of government by crediting rulers' decisions with the force of 'the will of the people'. The shorthand use of 'America' or 'the Russians' for their governments, clearly exaggerates the executives' moral significance. The notion of the variety of people in any compass of territory having a unity of objectives and attitudes and somehow transmitting this for voice and action to one individual is obvious nonsense. Arrow has demonstrated the impossibility of compiling by vote an acceptable collective ordering of society's desires, acceptability being defined by principles which exclude discrimination, dictatorship and the dependence of choice on irrelevant alternatives. Any resort to more indirect ways of gathering the will of the people together to be exerted by one person are obviously working in the realms of magic. Historically, the interests served by the nation state have been those of groups who amass sufficient political clout to force their wishes on others. The more powerful, not always the majority, thrust the desires of the weaker aside.

The realities of politics both within and between nations are obfuscated by talk of territories having motives and by imputing to rulers the ability to express the uniform collective will of the population of a territory. This mode of expression also gives a false impression of continuity of interests and purpose. The territory may not change, but the leadership, the population and their attitudes and interests do — continuously. To identify the piece of land occupied by this changing population as the actor in international affairs, even as a matter of convenience of expression, builds unnecessary rigidities into our view of the world. This is especially so where some doctrinal position is involved. There have been long-standing policy positions which survived in the annals of diplomacy, but these represented the agenda of some powerful clique which established a continuity of power and interests by institutional indoctrination.

There is no doubt that the myth of nationality and the identification of territory, people and the state, is a most powerful force in history. It has led millions to participate in or consent to extremes of altruistic sacrifice and savagery. Masses of people have been imbued with the belief that the source of their spiritual and material well-being is the nation, to which they owe their ultimate loyalty. The numbers involved do not, however, disprove the case that the myth is built on sorcery and not objective geographical reality.

Whatever the mysteries which justify political authority, most polities' arrangements do vest sovereign power to change the law and wage war in the person of an individual head of state. The effect of this is to place the fate of the world in the hands of very few people. The problems of perception and identity associated with political authority have become matters of life and death for us all. Whereas it might have been possible to find a refuge to avoid being drawn into the turmoil of war in the past, nuclear war may now visit us all, whether we are in a combatant nation or not.

The degree of concentration of destructive might brought about by nuclear power very literally puts the lives of nations in the hands of the very few who command the nuclear arsenal and at the mercy of their attitudes and motives. By contrast with the mythical notion of the collective vitality of the nation, the prospect of collective death is now very real. If those who hold this power of life and death have confused and simple-minded notions of the nature of peoples, nations and states and of their own function, so much the worse for us all. The greatest danger is that the complexities of relations between nations may be reduced in the minds of those in command to the simplicity of a duel with the last one left standing as 'the winner'. The Manichean division of people, babies and all, into good and evil camps depending on what territory they live in, holds the prospect of the end of mankind, for the heat and dust from such a duel will choke all habitable parts of the earth.

The image which captures the ultimate choice of statesmanship is that of 'the button', the trigger of nuclear war. In a recent investigation of this mechanism, Ford dispelled any comfort we might have drawn from the notion of a well-informed, reliable, defensive, fail-safe machine with which the president commands the nuclear forces of the USA. Ford presents evidence of a disparity between the public announcements of a retaliatory, deterrent intent and the practical war plans of the US military. 'Declaratory policy' for public consumption is that the USA will never make a pre-emptive first strike. It seems, however, that since 1954 and the days of General Curtis LeMay, the 'anticipated operational plan' calls for the US to attack first in the case of an emergency. The current Single Integrated Operational Plan (SIOP), which details possible responses to crises, does include the option of attacking targets in the USSR with a surprise strike seeking to kill the political and military leadership, decapitate the command system and thus prevent retaliation. Soviet strategists have always been open about the advantages they see in a first strike aimed at US nuclear weapons and their command system.

Given recognition of this strategic option, it would not be surprising if discretion over if and when to retaliate were not diffused through the military hierarchy by delegation. Submarine commanders with tenuous communications with the rest of the world can fire nuclear weapons without hindrance from higher authority. The general flying in the 'Looking Glass' command plane could unilaterally order an attack. This post has no link to the USSR's leadership for negotiating an end to any exchange. If it were decided to make a first strike, a mere phone call from the president would set things in motion. The military has control of the triggers. The president functions as a safety-catch, which could conceivably be overridden. The ultimate seat of decision for the nation state, then, rests in the upper echelon of a self-perpetuating military caste. The ethos of their profession is warlike. Its members are indoctrinated in the spirit of fighting to win at all costs. The geographic consciousness of general officers is educated to a strategic level of resolution which reduces the details of our occupation of the land to a gameboard-like format for the great confrontation.

Readings

The best and most far-reaching analysis of political motives runs as a strand through:

McNeill, W.H. *The Rise of the West* (University of Chicago Press, Chicago, 1963)

This is drawn out and elaborated in:

McNeill, W.H. *The Pursuit of Power: Technology, Armed Force and Society since AD 1000* (University of Chicago Press, Chicago, 1982)

The foreign policy of the USA and its relations with the USSR are widely chronicled and analysed. Some sources used for this chapter include:

Ambrose, S.E. *Rise to Globalism: American Foreign Policy 1938–1980* (Penguin, New York, 1980)
Barnet, R.M. 'The Annals of Diplomacy: Alliance I and II', *The New Yorker*, 10 and 17 October 1983, pp. 53–105 and 94–167
Melanson, R.A. (ed.) *Neither Cold War nor Detente: Soviet-American Relations in the 1980s* (University Press of Virginia, Charlottesville, 1982)
Nathan, J.A., and J.K. Oliver *United States Foreign Policy and World Order* (Little, Brown, Boston, 1976)
Northedge, F.S. (ed.) *The Foreign Policies of the Powers* (Praeger, New York, 1969)
Schlesinger, A. 'Foreign Policy and the American Character', *Foreign Affairs*, vol. 26, no. 1 (1983), pp. 2–16
Steel, R. *Pax Americana* (Penguin, New York, 1977)
Ulam, A. *The Rivals: America and Russia since World War II* (Penguin, New York, 1976)

The 'Impossiblity Theorem' is proven in:

Arrow, K.J. *Social Choice and Individual Values* (Wiley, New York, 1951)

The nuclear trigger is investigated in:

Ford, D. 'The Button', *The New Yorker*, 1 and 8 April 1985

3 VIEWS OF THE WORLD

Images

Knowledge begins in sensing simple patterns in the tangled web of things and events of existence. The next step is making general rules out of patterns to make life less uncertain and easier. By adjusting the depth and sharpness of our senses we can pick out from the noises, sights, smells and textures of the world the main elements and circumstances which lead to serving our immediate purposes. We seek a simple, comprehensible pattern which relates our desires or fears to components of the landscape over which we can exert some control or which we can detect more readily. Science consists of generalising such perceived relationships and expressing them in the form of an instrument or the set of symbols whose workings are known to us. By manipulating the instrument or the language it is possible to come up with a prediction or prescription to inform our action in any particular situation. The observations of reality, and thus the form of the predictive device, are usually made with some particular purpose in mind. They will not necessarily be effective aids to choice where we are faced with a different scale and scope of events. Methods of prediction or prescription will certainly become outmoded as the tools and equipment with which we seek our ends are made bigger and more powerful. Models and theories have to be recalibrated or reformulated to address our need for knowledge at a different level.

By contrast with this usual purpose of theory to clarify the nature of events, when we consider generalisations on the geography of politics there is a suspicion that they might have been devised to obscure reality. Whether the purpose be to deaden a sense of guilt or to present a pseudo-scientific justification to the world, the effect on the gap between image and truth in geopolitical writing and utterances is similar. Naked aggression is clothed with fuzzy maps of territorial imperatives. The pursuit of power involves people in acts which most societies customarily condemn. Some circumstances have, however, been used to warrant killing, destruction and bullying. The conditions cited have often been some geographical criteria. At the crudest, differences between people in different places have been used to distinguish a superior 'us' from an inferior 'them', who do not deserve the territory they occupy or know

how to use it properly. Geopolitical generalisations have usually divided the world into two camps seen from the vantage-point of the self-
proclaimed moral superior of the two. The usual implication is that the
other is inferior or unworthy or bad. In some instances shades of
inferiority or unworthiness have been recognised. If this is the usual form
of world picture of an established order faced with the potential threat
of outsiders, the radicals who seek to gain power from change present
only a slightly different view. Revolutionaries have identified places from
which to spread change in politics and society. The territorial divisions
of their maps of the world are in terms of whether potential converts
or the obdurate hold the upper hand, thus separating good from bad.

None of these generalities could be deemed objective truth. The
obscurity and obliquity of the terms used have provided a means of talk-
ing about domination, suppression, ambitions and fears with minimum
offence to the sensibilities. The geographical images of world politics
which have been drawn are important not because they objectively explain reality, but because they interpret or express the intentions of certain
powerful sets of people. In some cases geopolitical ideas have fed the
mood of the time by providing an excuse for exploiting opportunities
or fuelling fear with mythical sub-human or super-human bogeymen.
In other times and places the geopolitical picture was drawn after the
event, providing a euphemism to justify bygone actions and the present
state of affairs.

Up to the sixteenth century, geography was confined by a lack of
global knowledge. As the earth's surface was mapped along with the
extension of European empires, descriptions of the geography of politics
reflected this perspective on the world. The hopes and fears of colonial
powers were the chief inspiration. Since the 1900s, the competition for
global hegemony between Great Britain, Germany, the USA, Russia and
China has been the principal inspiration of maps of world politics.

Traditional and Classical Views

We can catch only glimpses of ancient sentiments about political
geography through the limited supply of written remnants. However,
the notions expressed by Greek writers, for example, are probably
representative of a universal tendency to belittle other people elsewhere,
especially if it may become desirable to kill some of them. For the Greeks,
barbarian political and moral inferiority was bred of the climate and terrain of their homelands. Asians and others were soft and slavish because

they dwelt in a warm climate on level land with deep soil. Greek superiority arose from their hardy mountain setting. Temperate-zone, Greek hill-folks were obviously designed to dominate lesser breeds from torrid or frigid regions.

Another accessible, articulate ancient view of the world comes from the plains of the east. The Chinese thought in terms of a series of concentric, fuzzily defined zones of diminishing significance focused on agricultural China proper. Beyond this core lay a band, including Manchuria, Mongolia, Sinkiang, Tibet and Annam, for which the Chinese felt a strong affinity. The next band of inner Asia included Southern Siberia, Central Asia, Northern India, Indochina and Malaya. An outer Asian band represented the limit of trade and diplomatic relations, sweeping from the Caspian to the Persian Gulf, along the southern edge of Asia to the outer fringes of Indonesia, the Philippines and Japan. China had never exercised direct control over these lands. Beyond these lay the alien, outer world.

The view of the Roman elite, mapped by Pliny the Elder, was similar in structure. The seaways and shores of the Mediterranean provided an inner ring around the Roman core. From the ports of the Mediterranean, roads carried the power of Rome through an outer ring to the limits of the empire on the banks of the Rhine, Danube, Euphrates and Nile. The world picture of medieval Christendom shifted the focus to Jerusalem in 'T in O' maps: the Mediterranean provided the stem of the 'T' and the Black Sea, Aegean and Nile were the crosspiece. Asia and the site of paradise was at the top, the Atlantic at the bottom with Africa to the right and Europe to the left. This mapped a religious landscape. The contemporary *portolani* of the Mediterranean seaways were for steering ships rather than seeking salvation and they conformed more closely with material geometry. The Islamic worldview also continued to centre on the Mediterranean, with a world now divided between Christendom and Islam along the axis of the sea.

The Imperialist World

With the Iberian oceanic outflanking of Islam which had reached around Africa and across the Atlantic by 1500, geographic information began to accumulate rapidly to European sailors gaining control of the ocean basins and their margins. By 1600 Mercator and his successors had provided a framework for the presentation and use of these data and what came to be a widely disseminated image of the earth. By 1900 the sailors' continental outlines had been filled in, although the exploration of the surface continued until the 1960s. Indeed, by 1800 the contents of the continents were well enough known for von Humbolt and Ritter to found modern geography with a global framework. In the 1850s Guyot was presenting Ritter's worldview to American audiences, amended to centre on the North Atlantic basin and imply a shift of the focus of power to North America from Europe.

In 1890 Mahan provided a strategic appreciation of the global situation and indicated how Manifest Destiny could be carried beyond the shores of North America. He pointed out that Great Britain's dominating role depended on sea power. The implication was that if the leaders of the USA wished to follow suit, they would have to acquire a string of naval stations from the Caribbean across the Pacific, complementing the Royal Navy's command of the Atlantic and Indian Oceans. Mahan's campaign and influence on opinion ran closely parallel to the American acquisition of suitably located colonies.

Around 1900 Ratzel presented a coherent pattern for political geography, emphasising space as the basis of power, which led him to look at North America, Russian Asia, Australia and South America as the future powerhouses of history. It took Halford Mackinder, however, to marry the significance of space and strategic location with the dichotomy between land and sea power in a catchy, simple map. Although Mackinder was distinctly British and imperialist in his sentiments and purpose, he did stand back far enough historically and geographically to place his 'pivot of history' elsewhere. He placed this pivot in the interior of Asia, with Britain on the maritime periphery of the picture. Mackinder sought in word and action to maintain the supremacy of Britain controlling the rim around the heartland. But changes were afoot in the heartland and the Americas which were to erode the Royal Navy's monopoly.

The World of Class Conflict

Like Adam Smith writing a century before them, Marx and Engels were not aroused much by geography. The beginnings of enquiry into the workings of industrial society were concerned with the question of the allotment of wealth between owners of property, the business class and labourers. Although Smith showed a keen sense of the economic significance of distance, he turned to geographical variations only briefly in his discussions of comparative advantage for agricultural production and defence expenditures and the relative merits of hunters, herdsmen, farmers and townsmen as soldiers. The British tradition of political economy then divorced itself from the heritage of the map-making of Sir William Petty and abstracted space and geography from its analysis. In a small island with comparatively cheap river and coastal navigations, this was a reasonable simplification to come to grips with the main issues with the least clutter of circumstances in the argument. When geographical problems were confronted they were treated categorically, as in David Ricardo's treatment of land rent as a function of soil quality rather than as having some continuous variation with distance. In treating trade, since ocean transport was so cheap per mile, distance-related costs could be held constant and thus dropped from the analysis without great loss of generality. Meanwhile in Mecklenburg, writing from a more managerial perspective, von Thünen's spatial insights were lost on the mainstream of classical economics. Marx and Engels were very much of this mainstream, introducing an Hegelian dynamic into an aspatial, class-structured model of society, geared to the mechanics of production. They foresaw revolution stemming from the urban workers, with their inspiration making Germany the most likely venue. Rural Russia was for them the least likely starter and the last participant in Europe's brave new world.

When it came to the action, the man who came out at the head of the revolutionary operation, Lenin, was a talented opportunist. He found himself in the saddle of a largely rural, Asiatic empire in chaos. To legitimise leading the communist revolution from this position, it was necessary to rewrite the simple geographic specifications of Marxist gospel, as well as its constitutional ones. As circumstances changed, so did Lenin's geopolitical position — each change being designed so as to set Lenin at the head of whatever was developing. In Russia in 1904 he saw himself as the leader of an insurrection sweeping from Russia into Germany to take charge of the world revolution at the centre of events. In Switzerland in 1914 Lenin was preparing himself to lead the revolution in Germany. This social explosion would be the synthesis resulting from the thesis of the German imperialist war for *Lebensraum* generating its antithesis of revolutionary conditions. Betting on a German revolution, Lenin devised the theory of 'socialism in a single country'. In 1915 he put aside notions of a United States of Europe. The geographical inequality of industrial devlopment meant that only Germany was ripe for socialism. This new ambition was quite disturbed in 1917 when the German foreign office offered to ship him back to Russia to foment insurrection there. But, faced with new prospects, he pieced together a new theory with a place for Russian revolution as a starting-point in the grand design. However, he continued to hope for a German revolution. The Treaty of Brest-Litovsk bought time for the Russian revolution with territory, but it also maintained a channel of communication open to Germany and Lenin continued to hope to transfer his seat of power to Berlin.

The realities of the power struggle in the Russian empire led Lenin to depend on non-Russian nationalities, although he remained a Great Russian imperialist at heart. The revolution had been given teeth by Latvian riflemen who formed the core of the Red Army. Lenin needed non-Russians to kill the large number of Russians who would have to be removed if he was to achieve success. The foremost of his non-Russians was Stalin, his 'amazing Georgian', who was the executive arm for a distinctly Asian turn in Lenin's policies. When revolution did not materialise in Germany a new worldview and theory was called for and by the end of 1919 the Russian empire in Asia became Lenin's new base for an attack on the empires of capitalism. Russia became *the* base for revolutionary leadership which took the form of 'the dictatorship of the proletariat', meaning the manipulation of the rural masses by an oligarchy with their power base among the industrial workers. There was no Marxist scriptural precedent for a revolution that began in the backward East with a strongly nationalist flavour. Lenin nevertheless proclaimed it the new base for world revolution. The British overseas realm was to be infested by missionaries sent from Turkmenistan into Iran, Afghanistan and India. In the West Lenin cautioned communists to do what they could for Russia within the limits of the law and live by the slogan of 'peaceful coexistence'. The nationalist spririt which Lenin harnessed to overthrow the opposition was offered the prospect of 'self-determination'. In practice this meant a fleeting period of official independence followed by annexation into a Union of Soviet Socialist Republics controlled by a Great Russian elite and its henchmen. In 1922, on his deathbed, Lenin had a falling-out with Stalin because the Georgian was violently short-circuiting annexation in his own homeland, thereby upsetting Lenin's image which called for a show of *de jure* independence first.

The ostensible issue of principle in the struggle for power between Stalin and Trotsky after Lenin's death was geopolitical. Whether the positions were intellectually derived matters of belief or whether they were two obvious extremes to take in a fight where Stalin was in the role of the organiser and Trotsky of the idealist, is hard to say. There is some evidence that both men had shifted ground on the questions involved before they set about destroying the other man's position. Trotsky finally took the stance that the first priority was international revolution. This was to be fostered to weaken capitalist governments so as to prevent them turning on Russia. Socialism in Russia was hardly feasible otherwise. Russia's role was to lead a permanent revolution, bringing proletarian power to Germany and Britain. These industrial countries would then provide Russia with manufactures in return for raw materials. Stalin's counterslogan with which to beat Trotsky was 'socialism in one country'. This offered the hope to Russians that they would not have to wait for the rest and sought to exploit Russian pride that they had succeeded in revolution before the advanced nations of Western Europe.

In 1927 Trotsky's bid for leadership failed and he was expelled from the Communist party. In 1928 Stalin launched the programme to achieve soicalism in one country. This consisted of the first five-year plan for industrialisation and the savage collectivisation of farming to feed this process. Without industrial production to arm its forces, Russia could not be defended against its enemies. Stalin was impressed by Russia's long record of vulnerability. More recently, intervention by Western governments in 1918–22 was evidence that the capitalists were out to destroy the revolution. Their gestures were not strong enough to affect the outcome, but they certainly prolonged the agony and deeply embedded the fear of capitalist destructiveness in the collective memory of the Soviet elite and people. In 1930 Stalin confirmed Russia's isolation with the defensive doctrine of capitalist encirclement. Quite obviously socialism in one country implies a surrounding host of hostile, capitalist imperialists. Stalin seems to have believed that war between the two camps was inevitable, but must be staved off by diplomacy to prevent the formation of an anti-Soviet coalition until Russia was strong enough. Talk of collective security and disarmament were obvious ploys to this end.

Geopolitik

Although Mackinder's worldview had a British inspiration, in identifying the source of danger for the maritime empires it did provide the prospect of power and domination for those who controlled this source. In 1904 Mackinder pointed to the north central plains of Eurasia as the pivot of history. These were impervious to navies but open to the sweep of cavalry or the thrust of railways. From this territorial base, Russia, whatever its constitution or economic arrangements, exerted the primary political pressure on the globe. This was felt in the five peripheral regions of Eurasia, East Asia, South Asia, the Middle East and Europe. Naval power could be brought to bear on this marginal crescent. After the First World War Mackinder amended this map and extended the pivot, calling it the 'heartland'. This was fringed by the inner crescent named above and then by an outer crescent consisting of the British Isles, Africa south of the Sahara, and Japan. The fundamental opposition of the world island of Eurasia and Africa sets the land power of the heartland off against the sea power of the crescents. Mackinder's purpose in constructing this map was to warn British politicians of the threat to Britain's maritime empire of the combination of Germany and Russia. By 1919 Mackinder had shifted the locus of power and strife westwards into Eastern Europe. His warning was that whoever combined the Eurasian plains with Germany would rule the world island, and whoever ruled Eurasia and Africa ruled the earth. The Americans were inconsequential in Mackinder's theory of war and power.

The grandiose vantage point from which these maps were drawn did not fit the prosaic style of British political and intellectual circles. Its lofty vision did, however, excite some who had been raised in the Ratzel tradition of political geography. A Swede, Rudolf Kjellen, who invented the word 'geopolitics', was particularly taken by it. So too was a Bavarian soldier, Karl Haushofer. On his way to Japan in the 1900s, the latter had been disturbed by the omnipresence of the Royal Navy. What Mackinder had written to warn the rulers of this British global empire of ever-present danger, suggested at the same time the way to break its hold. This was for Germany to seek an accommodation with Russia. Haushofer was writing in this vein in 1913 and extending the axis to include Japan. In 1924 he founded the *Zeitschrift für Geopolitik,* a journal with German domination of world affairs as its editorial goal.

There is some question as to how important Haushofer was in forming Nazi policy, and thus how influential Mackinder's map of the world. Certainly Haushofer, whose family seat was just outside Munich, was involved in the formative stages of the National Socialist Party. Rudolf Hess had been his aide-de-camp in the war. He met Hitler on several occasions while he was dictating *Mein Kampf* to Hess, and this volume did lay out Hitler's agenda. Amongst its racial hatred, geopolitical notions are employed. There is some semblance of an underlying plan, first to obtain control of the heartland and then to destroy the naval power of the UK and USA. There is, however, some ambiguity over the latter objective. Hitler elsewhere showed evidence of wishing to avoid a final battle with British sea power, since he regarded the continued existence of the British empire as an indispensable part of the world order. Much of what might be taken to be derived from geopolitical imperatives — such as the impelling desire for *Lebensraum* and the wider and wider definition of *Deutschtum* that went with it — were fairly commonplace notions in the pan-Germanic mythology of the middle classes.

Whatever went on in Hitler's mind, it certainly seems to have been more strongly focused on Europe than either Haushofer's or Mackinder's analysis would have dictated. Whatever opportunities presented themselves elsewhere were but a means to consolidating the European New Order. Like others with a lust for power before him, Hitler used whatever slogans or devices were at hand to promote the headlong outward drive of his circle of potence. There is little evidence of a global, geopolitical blueprint derived from Mackinder's work underlying his ambition. His deepest emotion was hatred of Jews and Slavs and this owed nothing to geopolitics. Whatever geopolitical ideas he seized seem to have been a matter of convenience. For the New Order in Europe there was an economic plan, drawn up by Alfred Weber the location theorist. This designated an industrial core consisting of Germany, northern France, Bohemia, Moravia and northern Italy. This was surrounded by a belt of basically agricultural territory. If you add England, south Wales and south Scotland to the core and the rest to the periphery, this Nazi *Grosswirtshaftsraum* bears a strong resemblance to the image of European geography fostered by the officials of the EEC.

Hitler's decision to turn on Russia in 1941 clearly violated the precepts of *Geopolitik* as propounded by Haushofer and derived from Mackinder. Haushofer did not editorially voice any objection, but he did point out the difficulties that Napoleon and von Falkenhayn had faced in the vast spaces of the heartland. Clearly he was disappointed. In 1944 he was thrown into Dachau and his son was executed for his part in the plot against Hitler.

Whether it had an actual influence on Hitler or not, geopolitics was perceived by the opposition to be significant and Mackinder's works were dug up and digested. His terms, 'world island' and 'heartland', were used by statesmen and newsmen. The model was reworked to bring it up to date. In 1944 Spykman reintroduced Mahan's emphasis on sea power and modified Mackinder's map to produce a heartland and a rimland. The rimland, Mackinder's outer crescent, was the key to power. To prevent Germany dominating the world it was necessary for Anglo-American sea power to form an alliance with Soviet land power and prevent Hitler seizing control of all the Eurasian shoreline. With this done, future US policy should be directed to control the rimland or, at least, to exclude Russia from power over it.

Hegemonic Duopoly

Spykman thus gave early expression to what was to become the obsession of world politics in the second half of the twentieth century: the competition between the USA and USSR to control the fortunes of the world. The fundamental conflict for him was between Russia and the enveloping sea power for control of the rimland territories.

The American military also began their preparation for a possible Third World War with Russia in 1944. They bought the long-range B36 bomber even though it had no mission in the current conflict. There was a deepseated loathing of communism among the ruling classes of the West, which compounded an older fear and hatred of the Asiatic horde.

In Russia, faced with German pressure from the west seeking *Lebensraum,* Stalin had done what he might to hold off an inevitable battle while Russian forces were prepared. First, Foreign Secretary Litvinov made a pitch for collective security against Hitler's aggressive thrusting. Then his successor, Molotov, followed a policy of conciliation with Germany and Japan to turn their attention elsewhere while Russian industrial capacity was built up ready to meet them. Whatever its social and economic content, Stalin's policy emphasising isolation and encirclement by hostiles had the appearance of a reversion to Great Russian patriotism and imperialism. Even the missionary zeal of Marxist–Leninist cadres was not far removed in spirit from the messianic and pan-Slavic traditions of tsarist Russia.

When Hitler was defeated, what exercised Stalin immediately in 1945 was the need to create as broad and controlled a cordon of defence as possible to stymie any effort by the Western allies to overwhelm the USSR in its weakness during the aftermath of a devastating war. The industrial and military might of the USA was evident. There were more American soldiers than Soviet soldiers. The Soviets had a mere 60 divisions of battle-worn troops in Eastern Europe.

If Churchill and Roosevelt were willing to deal with Russia by negotiating post-war spheres of influence, after Roosevelt's death anti-Soviet attitudes hardened in the face of Stalin's intransigence over Eastern Europe. By 1947 this antipathy crystallised into the Truman Doctrine, the Marshall Plan and the H-bomb programme. The geographic expression which came to characterise this parting of the ways was uttered by Churchill in Fulton, Missouri, in March 1946: an 'iron curtain' had fallen across Europe. The analogy had been used previously by Queen Elizabeth of Belgium, drawing an iron curtain between herself and Germany in 1914.

In March 1947 the Truman Doctrine extended the conflict with communism from Greece and Turkey to a worldwide struggle between light and darkness. In June 1947 the Marshall Plan was announced with the objective of rebuilding a democratic Europe strong enough to stand against communist expansion. In July 1947 George Kennan provided a theoretical structure for the general feeling with 'containment' policy. Although Kennan denied any debt to geographic formulations deriving from Mackinder's heartland and its surrounds, he was certainly aware of them and his writing resounds with the need to encircle the threatening land power.

The Soviet interpretation was of an aggressive campaign against Eastern Europe. Stalin's reaction was to consolidate the Soviet hold there, starting with Czechoslovakia.

The sequence of collapse of the Kuomintang in China, culminating in the triumph of Mao Tse Tung in 1949, roused American leaders' concern over the Pacific rimland once more. In 1950 Secretary of State Dean Acheson described an American defence perimeter in the Pacific running from the Aleutians to the Philippines, excluding South Korea and Formosa. This exercise and subsequent events showed that it was not easy to draw lines of containment and delimit spheres of influence in a cut-and-dried fashion. In the space between the powers there was room for manoeuvre and indirect competition. Back in 1915 Fairgreve had used the term 'crush zone' to describe the belt of small countries lying between the heartland and the sea powers. In 1963 Cohen presented what he saw as the geopolitical equilibrium consisting of core areas divided by shatterbelts. With the Soviet acquisition of a nuclear arsenal after 1949, the temptation for full-frontal opposition was diminished and since that time the conflict between the hegemonic powers of the USA and USSR has mostly taken place indirectly in the shatterbelts.

On the American side, fear of militant, missionary communism was broadcast in 1947 by William Bullit in an article in *Life* magazine which conjured up a monolithic power centred in Moscow, spilling communism outward and engulfing the world via China and Southeast Asia. This ideological flow was compounded by the creation of some historical myths. In 1952 Morrison wrote of an 'urge to the sea' as the single geographical imperative which had driven Russian foreign policy since the days of Peter the Great. In 1954 Wiens tied the spread of communism in with the ethnic imperialism of the Han Chinese in a new version of the 'yellow peril' myth. Later, when courting the USA, Brezhnev appealed to this base racial fear by referring to 'we Europeans', as opposed to the Asiatics. In the 1950s, however, whatever its driving force, the surge of fear was of the red peril, of a combination of Russia and China. The prospect of the red tide of communism was so horrifying to Secretary Dulles that he talked of carrying war into the enemy camp with phrases like 'roll back' and 'liberation'. In practice, however, his policy was one of containment, intervening in Southeast Asia and completing a chain of air bases encircling the USSR and China.

In a 1953 meeting of the Joint Chiefs of Staff, Admiral Arthur Radford proposed to relieve the French Foreign Legion in Dien Bien Phu with a nuclear bombing strike against the Viet Minh. To justify such drastic action, he likened the loss of nations to the communist camp to the chain reaction of a collapsing row of dominoes. This analogy took Eisenhower's fancy and the catchy phrase 'like a row of falling dominoes' entered into the official vocabulary and the language of the media. Mackinder had used a similar image, a row of ninepins, in 1934 when writing of the fall of small nations to the heartland state. The mental map of the world that went with dominoes seems to have persisted with a duality of centre and periphery such as Mackinder had first suggested in 1904. The domino simile was taken up by Kennedy and began to be referred to as a theory. Nixon and Ford subscribed to it, as did Kissinger and now Reagan. This geographical spectre has haunted the words and deliberations of American makers of foreign policy and provided a glib catchphrase for journalists for 30 years.

In the Soviet Union, Stalin was finally repudiated in 1956, three years after his death. Khrushchev gained the upper hand and brought a more expansively aggressive worldview to bear. The Communist Party elite had retained a faith in the inevitability of revolution and the USSR's mission to undermine the international capitalist structure. There was an inner consistency to the seemingly discordant policy which Khrushchev pursued of expanding Soviet influence in the Middle East, Africa and Latin America while at the same time seeking 'coexistence' with the USA. Khrushchev sought to expand the influence of the USSR where it would embarrass the Western alliance, and to move into power vacuums. 'Wars of national liberation' around the globe provided the occasion to inject a Soviet presence and outflank the USA. The Soviet government did not generate these wars but was not loath to exploit them and play upon the traditional hostility directed against the formal, colonial empires of Europe by the US establishment. This anti-colonial obsession could be used to open rifts between America and Western Europe. Stalin's doctrine of capitalist encirclement was dismissed. Khrushchev believed that the USSR with its nuclear arms was on the brink of economic and military superiority. He thrust the Soviet presence onto the global scene and

posed the question: who was surrounding whom? He was willing for
Russia to coexist with the prolonged decay of the decadent West,
gradually assuming a wider sphere of dominance by indirect intervention.

The US government responded to this new aggressive spirit by
extending the area of containment of the Truman Doctrine of 1947. In
1957 the Eisenhower Doctrine took in the Middle East. There was little
prospect of confrontation, however, since Moscow was not fomenting
political instability but merely exploiting existing situations.

Third World Revolution

Khrushchev's departures from Mao's interpretation of the orthodox party
line provided the occasion for a rift to open between the USSR and China
by 1963. This brought the map of international loves and hates back to
a more traditional pattern. Denouncing coexistence as a renunciation of
revolutionary armed struggle, the Chinese leadership produced a new
geopolitical view of the world. A three-way struggle for the earth was
under way between the true revolutionary spirit embodied in China, the
revisionists in Moscow and the imperialist reactionaries led by the government of the USA. Around these foci, the rest of the nations were arranged
in two intermediate zones. The first consisted of the peasant nations of
Asia, Africa and Latin America. The second included industrialised
Europe and Japan. Chinese attention was concentrated on the first zone
which was referred to as 'the countryside of the world'. Mao had conceived of a Chinese revolution based on the peasantry before 1927. By
1939 he had proclaimed that China's example in this would be followed
in all colonial countries. The most important imperative force in history
was the antithetical clash of peasant violence directed against the colonial
and neo-colonial powers in Asia, Africa and Latin America. Any
insurrectionary movement, be it nationalist, tribal or economic, was
worth assisting if it helped to destroy the existing imperial structure.
After a first wave of nationalist or tribal revolutions, party cadres would
provide a second wave to displace inept nationalist elites, still heavily
dependent on the imperialist world. A mythical version of Chinese history
between 1911 and 1949, with the revolutionary peasantry under Mao
surrounding and overwhelming the cities, was blown up to global proportions.

This new global vision was given public expression by Lin Piao in
his pamphlet *Long Live the Victory of the People's War*, published in
1965. North America and Western Europe, cast as the 'cities of the

world', were surrounded by the 'countryside' of Asia, Africa and Latin America. The failure of revolution to transpire in the capitalist world transferred the honour of revolutionary leadership to the country folk, who were two-thirds of the world's population. These would overwhelm US imperialism and its lackeys, including the Soviet revisionists who urged gradualism and peaceful coexistence out of fear of nuclear weapons. There was no need of such timidity for the true revolutionary because the imperialists would not use such weapons against people's wars out of fear of universal condemnation. Mao was willing to contemplate taking on a nuclear power and overwhelming it by sheer numbers of survivors. There have been a number of efforts to put this revolt of the countryside in train. The most self-conscious perhaps was Che Guevara's failure to establish a *'foco insurrecciónal'* in the Bolivian chaco.

Since the Chinese establishment flung out this challenge, the Soviet leadership has been drawn between appearing more militant than China or abandoning the leadership of those with violent revolutionary ambitions. The advantage of the second course is that it reduces the risk of a direct clash with the USA and attracts a following of the more sedate elements in the unaligned world. The temptation to dump their ideological baggage in favour of playing the Great Russian world power is strong. Although Brezhnev continued to intervene to sway the course of events far from the USSR in Africa, Asia and Latin America in order to disrupt the capitalist order, he limited action to fairly inconsequential places and mostly worked for stability where there was a prospect of a head-on clash with the USA — in the Middle East, for example. His intervention in Afghanistan was the action of a Russian imperialist. This operation evidently ran counter to the advice of his successor, Yuri Andropov. Andropov's power base, in the upper ranks of the KGB, has displayed quite a cautious, nationalist attitude to the rest of the world for some time. Mikhail Gorbachev has displayed no signs of a radical departure and seems intent on winning European and Japanese neutrality.

In China since the death of Mao and his demotion from deity, there has been a distinct curtailment of militant tendencies and, on the face of it, a return to the traditional Sinocentric view of the world. In this schema, geographic distance from the borders of China, rather than political distance from the epicentre of the authentic revolutionary force, is more significant in determining importance.

Through these changes in the communist world, the attitudes of Western leaders seem to have remained impressed with a Mackinder-like view of the world. A violently revolutionary heartland is surrounded by a rim of nations whose loyalty must be won to contain the ever-present red peril to social stability. This has not been laid out as a blueprint for action, but it is the obvious interpretation of foreign policy utterances and actions. The continuing use of the domino motif is an expression of this frame of mind. Kennedy subscribed to this 'theory', as did Nixon and Kissinger. Carter bucked the trend of geopolitical simplification in favour of moral naivety, until Brzezinski established his hold on foreign affairs in the administration. In 1978 Brzezinski made a grand geographic sweep and put a series of danger points and potential threats to the oil supplies of Europe, the USA and Japan into an 'arc of crisis'. This curved through North Africa and Southwest Asia from the Horn of Africa to Chittagong. In 1980 Carter declared that the Persian Gulf region was of 'vital interest' to the USA and that he would counter any outside effort to gain control of territory here with force. Reagan's view of the world was formed in the 1950s and he talks as a militant cold warrior and seems to believe that he has a mission to overturn communism. Members of his administration have stated an ambition to overthrow the Sandinista government in Nicaragua, for example. Even Mexico has been referred to as a domino, much to its foreign minister's chagrin. One form or another of a world of good guys and bad guys with a buffer zone to be competed for in a contest of strength, influence and resolution, seems to lie behind the pronouncements of those who have the ear of the decision-makers. Beneath Jeane Kirkpatrick's directional confusion of 'an eastern front on our southern flank' with reference to Russia and El Salvador, there seemed to be a map of two camps with a contested band of pink between the red and white. Kirkpatrick's efforts to direct attention to Central America were not put in terms of the balance of moral issues and social and economic circumstances there, or any local definition of US interests, but of a global struggle. Her views were endorsed by Kissinger in *Time* (9 May 1983), where he is quoted as having said: 'It is time we stopped arguing only about how much democracy there is in El Salvador and began to understand America's strategic interests are at stake.' The Reagan Doctrine proclaimed the objective of not merely containing but of rolling back limits of 'Soviet expansionism' by supporting insurgents wherever they seek to overthrow governments which are friendly to the USSR. It is noteworthy, however, that as she stepped down from the administration in December 1984, Kirkpatrick issued a renunciation of the 'superpower rivalry model of the world'. She admitted that most of the world's problems are regional conflicts having little to do with Moscow or Washington.

It does seem that the rulers in Moscow and Peking have returned to in more traditional, nationalistic view of the world. There may be some residuum of revolutionary zeal in their corridors of power which throws another source of ambiguity into a decision process clouded by the uncertainties of a complex world; as things stand, though, there is little indication that a consistent over-all Soviet foreign policy strategy exists apart from the neutralisation of Western Europe and Japan. Meanwhile, the leadership of the USA seems to have adapted a simplified ideological viewpoint which derives in part from an excessively simple mental map of political geography.

Readings

In addition to those works on the rivalry of the USA and USSR referred to at the end of the last chapter, this one was built on some standard works on political geography:

Blij, H.de *Systematic Political Geography* (Wiley, New York, 1967)
Pounds, N.J. *Political Geography* (McGraw-Hill, New York, 1972)

The imperial view is presented in:

Mackinder, H.J. 'The Geographical Pivot of History', *Geographical Journal*, vol. 25 (1904), pp. 421–44
——— *Democratic Ideals and Reality* (Constable, London, 1919)
Mahan, A.T. *The Influence of Sea Power upon History 1660–1783* (Little, Brown, Boston, 1900)
Parker, W.H. *Mackinder: Geography as an Aid to Statecraft* (Clarendon Press, Oxford, 1982)

The practice of communist geopolitics is dealt with in:

Lawrence, J. *A History of Russia* (Grove Press, New York, 1961)
Page, S.W. *The Geopolitics of Leninism* (Columbia University Press, New York, 1982)
Sprout, H. and M. *Foundations of National Power* (Princeton University Press, Princeton, NJ, 1945)

Geopolitik is chronicled in:

Dorpalen, A. *The World of General Haushofer: Geopolitics in Action* (Farrar & Rinehart, New York, 1942)
Whittlesey, D. *German Strategy of World Conquest* (Farrar & Rinehart, New York, 1942)

The train of events and attitudes since the Second World War can be traced through:

Bullit, W. 'A Report to the American People on China', *Life*, 13 October 1947
Cohen, S.B. *Geography and Politics in a World Divided* (Random House, New York, 1963)
Kennan, G.F. (X) 'The Sources of Soviet Conduct', *Foreign Affairs*, vol. 25 (July 1947), pp. 566–82
Morrison, J.A. 'Russia and Warm Water', *United States Naval Institute Proceedings*, vol. 78 (1952), pp. 1169–79
Spykman, N. *The Geography of the Peace* (Harcourt, Brace, New York, 1944)
Wiens, H.J. *China's March towards the Tropics* (Shoe String Press, Hamden, Conn., 1956)

The domino theme is traced through:

Asprey, R.B. *War in the Shadows* (Doubleday, New York, 1975)

The significance of China emerges through:

Ginsberg, N. 'On the Chinese Perception of World Order', in Tang Tsou (ed.), *China's Policies in Asia and America's Alternatives*, vol. 2 (University of Chicago Press, Chicago, 1968)
Griffith, W.E. *The Sino-Soviet Rift* (MIT Press, Cambridge, Mass., 1964)
Lin Piao, *Long Live the Victory of the People's War* (Beijing, 1965)

Projections of Soviet attitudes are to be found in:

The International Institute for Strategic Studies 'Prospects of Soviet Power in the 1980s', *Adelphi Papers*, nos 151 and 152 (1979)

4 THEORIES

Choices or Space

I do not intend to survey what has been written in the way of theories of international relations. This has been done recently and fairly completely elsewhere (see books listed at the end of Chapter 1). For present purposes it is more useful to look at two different types of theoretical approach to the relations between states.

On the one hand, there is an approach which tries to retain as rich a representation of the decision-making process as possible. The social order of a state and how it changes are important influences on policy. Complex domestic conditions come to bear on the choices made in the competition for influence and resources between opponents with similar problems. To come to grips with these aspects of the problem, models deriving from economic theory have been adapted to international affairs. When matters get down in the final analysis to the strategies employed by the contestants in a conflict, game theory comes into play. For such theories, geography is merely a complicating factor and is usually assumed away.

By contrast with these formulations, in order to say anything interesting about space and location in the contest between nations, it has proved necessary to sacrifice details of the decision calculus and of conflict-resolution in favour of a spare setting for the exercise of mathematical rigour.

Economic Models

There are two main streams of opinion on the object of foreign policy. The classical, realist view is that security and power are the ends of statecraft. There is a more recently founded school of thought that puts the well-being of the population of the state as the prime goal of policy. This sets aside Plato's maxim that politics should not be housekeeping written big. Domestic stability and the prosperity of the people become the chief objectives of statesmen. To reconcile these positions the state can be treated in similar fashion to the firm in the theory of production. The state becomes an organisation for the provision of protection and welfare in return for revenue, competing with other states for influence and resources. The statesman becomes an automaton, like the entrepreneur of classical economic theory, who responds to situations so as to get the most of an objective which is set down as desirable from the start. Decisions then can be described by movements over a mathematical function representing the objective, towards a peak on a hill of desire. Where there are undesirable outcomes that go along with getting what you want, then the best choice rests where there is the greatest difference between good and bad outcomes, where the net good is greatest.

In the mass, these decisions made by many producers in the light of their own circumstances can be represented by movements along a supply curve. The state of rest of the system is found where this intercepts a curve describing the mass behaviour of buyers, a demand curve. From all other positions there will be a tendency towards this equilibrium, at a price where supply satisfies demand.

The adaptation of this model to politics turns statesmanship into management. Since leaders become managers, reacting mechanically to given, popular objectives, then, the monopoly of violence, which establishes authority at home and power abroad, is provided with the broadest base of legitimacy and is made popularly, if not morally, acceptable.

The people who command the state and make choices on everyone's behalf, are pictured as making corporate, hedonically rational decisions in their competition with other states for resources and territory. International politics becomes an exercise in net benefit maximisation. The costs of government and its beneficial outcomes could be described by two separate curves relating these quantities to some measure of the size of government. If the cost curve displays a form presumed to hold in productive enterprises, it will first decrease as economies of scale are realised and then, after a singular minimum, increase as diminishing returns to scale are brought on by congestion. If we presume that an extra bit of something good is more valuable when you have little of it than when you have a lot, then the benefit curve will slope down to the right, as in Figure 4.1. This is the curve of government revenue, and is thus measurable in money, as is the cost curve. The amount of government intervention which produces the greatest excess of benefit over cost is found at the intersection of these two curves, a. The surplus of benefit over cost is given by the area between the two curves where the benefit curve is higher than the cost curve, shaded in the diagram. At a this quantity is greatest. This is the scale of government towards

Figure 4.1: The Economics of Government

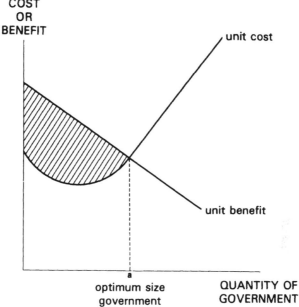

which statesmen could be expected either to expand or contract their operations and authority. If we introduced a third axis for the amount of territory within a state, the cost and benefit functions would become surfaces and their intersection at the planes of the axes would simultaneously determine the best size of state and amount of government within that territory. At the level of international interaction, this kind of construction could indicate desirable levels of trade and investment or military involvement, for example.

In particular, such a treatment of politics provides a structure to examine the emergence, rise and fall of the British empire, for example, or the hegemonies of the USA and USSR. The theory of competition among a few large producers for the market for a particular good offers an obvious inspiration for this.

The Greek classification of governments had oligarchy as the rule of a state by a few people. Faced with the realities of imperfection in markets, economists came up with theories of the behaviour of monopolists, duopolists and oligopolists, and their equivalents as buyers. The perfect market of price theory consisted of large numbers of buyers and sellers who could not individually influence price by their choice of how much to produce or consume. Real markets often had one, two or a few buyers or sellers who could sway prices by their individual decisions on quantities to offer or purchase, or who could set prices. We have not yet got to the stage where world politics is a monopoly, although there are deemed to be several sources of ambition to this end. The conditions of the 1950s, when the USA and USSR stood out above all others, could reasonably be called duopolistic competition. As far as nuclear might is concerned, this is still a fair approximation to the state of affairs. Outside of this terrible match, the 1960s saw the reassertion of independence by Mao and de Gaulle. In the 1970s Brandt began to seek greater self-reliance for West Germany with his *Ostpolitik*. By the end of the 1970s the Japanese leadership had gained enough economic confidence to strike its own postures in politics. World politics in the 1980s, then, will be more like oligopolistic competition, with control of the fortunes of the world contested between more than two, but few, powers. In this case the equipment of economic theory can be brought to bear on questions about the nature and stability of the outcome of competition between three or more powers for global dominance, as opposed to competition between two powers.

This resort to economics, however, only brings the question back full circle to strategies for games as a model. No simple result can be produced with the apparatus of cost and revenue curves determining the outcome of competition among two or a few sellers. In any reasonably realistic representation of the commercial world, which does not reduce enterprise to a knee-jerk response to one signal, there will always be a temptation to undercut a competitor and cut him out of the market altogether in order to gain monopoly power and reap its profits. There will always be fear that an existing competitor will do the same to you or that a new producer will set up to wipe both of you out or, at least, cut deeply into your shares of the market. Such circumstances may tempt sellers to agree among themselves on prices or quantities to reduce the uncertainties of competition. In the economic setting this is clearly against the interest of buyers, enabling sellers to reap and share the spoils of surplus profits with little risk. The outcome of competition comes to depend on decision-makers' expectations about their rivals' behaviour, which in turn depends on the rivals' reciprocal estimates of their actions in the future. Possible actions include not only aggressive competition but also collusion.

Game Theory

Conflicts in economics and also in politics can be viewed as games of strategy. In these games, events are not wholly matters of chance and outcomes are influenced by choice. The game is defined by the rules for playing, specifying moves, players, the order of play, the information available to each player and what ends a game. The result of a game is a distribution of payoffs to each player. The basic game is between two players, with one's gain being the other's loss, drawn from a fixed total payoff. The player's decision consists of the choice of a strategy. A strategy is a list of actions for every move of the game, one action for every possible move a rival can make. Each player may have a number of different strategies. The set of strategies for one player lists all possible ways of carrying out one play of the game, each strategy taking into account all possible responses by his rival.

The mathematical foundation for the analysis of such a model of events was laid in 1928 with von Neumann's proof of the minimax theorem. The relationships of payoffs to strategies can be represented by a system of linear inequalities. The conditions under which such a system can be solved involves expressing the relationship between the original (or primal) system and another system (its dual) which uses the coefficients of the primal inequalities to form new equations or inequalities. The minimax theorem states that the solution for the whole system is found where the primal and the dual meet at a minimax saddle-point. Here the maximum value for the primal set equals the minimum value for the dual. Translated into terms of a game of strategy between two rivals, with a fixed payoff to share between them, the best of the worst possible outcomes for one competitor will correspond to the lowest payoff his rival will allow him. The lowest payoff one allows the other determines, by subtraction from the fixed total, the greatest gain any player can guarantee himself. The maximum of the minimum amouts is the maximin. The minimum share left by its selection to a rival, establishes his maximum, the minimax.

This is most clearly understood by writing it out algebraically. For any strategy i chosen by a player A matched by strategy j chosen by player B, the payoff to A is written p_{ij}. If A has m strategies and B has n strategies, the payoffs can be arranged in an $m \times n$ table $P = |p_{ij}|$. The rows of this payoff matrix refer to A's strategies and the columns to those of B. Row i gives A's payoffs for strategy i, these being dependent on which strategy B uses. For a zero-sum game, the matrix P presents the payoffs to A as he tries to win as much as possible. B's payoffs are the negatives of A's, and B is portrayed as trying to keep A's winnings as small as possible.

Both players are assumed to know all the strategies available to themselves and their rivals. They are also assumed to be rational in the sense that A will always act so as to win as much as possible and B will do everything to thwart him in this.

If A uses strategy i, he is sure of getting the least value in that row

$$\min_{j} p_{ij}$$

no matter what B does. B will always allow him this. If A is shrewd, he will adopt a strategy which gives him the maximum of these minima

$$\max_{i} \min_{j} p_{ij}$$

B seeks to hold A's winnings to the least possible. If B uses strategy j, he is sure that A will not get more than

$$\max_{i} p_{ij}$$

whatever he does. The best strategy for B, then, would be the one which minimises B's maximum loss

$$\min_{j} \max_{i} p_{ij}$$

If there is a value in the payoff matrix p_{gk} such that

$$p_{gk} = \max_{i} \min_{j} p_{ij} = \min_{j} \max_{i} p_{ij}$$

then the game has a saddle-point. The best strategy for A is g and for B is k.

For circumstancess where no such saddle-point exists between two sets of pure strategies, the notions of chance and mixed strategies are introduced. Suppose that the players do not choose the pure strategy to use, but that it is selected by chance. This may be a matter of circumstances intervening to make some of the moves in the game or it may be a matter of a player's choosing his strategy by a chance device, the roll of dice or from a source of random numbers. For A chance selects strategy i with probability

$u_i \geq 0, \sum_m u_i = 1$

and for B another game of chance chooses strategy i with probability

$v_j \geq 0, \sum_n v_j = 1$

Chance then selects strategies, and players do not know what their own strategies nor those of their opponents will be, before the event. The vectors u and v define mixed strategies with u_i being the probability of selecting pure strategy i and v_j being the probability of selecting j. The outcomes of games are no longer sure. Players are faced with only expected payoffs.

If A employs a mixed strategy defined by u and B plays the strategy defined by v, then A's expected payoff is

$$E(u,v) = u'Pv = \sum_{i,j} u_i p_{ij} v_j$$

To solve the game the vectors u and v have to be found. A wants to find a u which maximises his expected winnings. For any u he selects, his expected payoff will be at least

$$\min_v E(u,v)$$

This is maximised over u

$$M_A^* = \max_u \min_v E(u,v)$$

Meanwhile, B can hold A's winnings to

$$M_B^* = \min_v \max_u E(u,v)$$

If there are vectors u*, v* such that

$$M_A^* = M_B^*$$

then there is a generalised saddle-point and the game has a solution. The expected winnings of A are M_A^*, and this is the value of the game.

Using the notion of duality, it can be proven that u* and v* always exist such that

$$M_A^* = \max_u \min_v E(u,v)$$

$$= \min_v \max_u E(u,v)$$

$$= M_B^*$$

$$u^* \geq 0, \, lu^* = 1, \, v^* \geq 0, \, lv^* = 1$$

The solution of a two-person, zero-sum game involves finding the mixed strategies described by u,v and the value of the game

$$M^* = M_A^* = M_B^*$$

This can be solved as a linear programme in which the extreme value on a linear function is found with respect to a bounding set of constraints. Such a solution exists for every n-person game with a finite number of strategies, but the solution need not be uniquely determined.

If it is not true that one player's gain is another's loss, then we have a non-zero-sum game and the scope for bluffing, feinting and colluding opens up, as in the game of Chicken, for example. In Chicken, there are two strategies available to each of the two players, chicken out or hang tough. A payoff matrix for this game is given in Figure 4.2. The first to chicken out loses with a payoff of $-\frac{1}{2}$, the winner getting $+\frac{1}{2}$. If both chicken out there is a draw with zero payoff for anyone. If neither gives in, the games ends in disaster with a payoff of -1 each. There is thus an incentive to make a deal to both chicken out and make a draw of it. But there is always a temptation to renege on the deal, hang tough and win. This applies equally to both players and in this dilemma lies the seed of destruction.

The ambiguity over deterrence or pre-emption in US nuclear strategy has provided the setting for the ultimate game, but it does not have a zero-sum outcome and few if any will be around to appreciate the toughness displayed.

Figure 4.2: Payoff Matrix for Chicken

	B	
	CHICKEN (A,B)	HANG TOUGH (A,B)
A CHICKEN	(0,0)	($-$ ½, ½)
HANG TOUGH	(½, $-$ ½)	($-$1, $-$1)

The rules of games where co-operation is a prospect allow for a discussion of strategies, with or without bribes (promises to transfer payoff) on the side. A co-operative solution can be determined for a two-person game with no side payments. When the number of players is expanded to three or more, mathematics will not yield a well-defined solution. When the number of players is pushed to the limit of infinity, however, fuzziness resolves into sharp-pointed results. But such conditions are hardly relevant to international affairs.

Game theory can cast analytic light on two-way conflicts. In the 1950s, for example, the RAND Corporation's analysts explored the use of increased military defensive resources in a game theory analysis. Intuition suggested to them that more defensive capacity should be used to extend coverage to hitherto unprotected targets. A game theoretic formulation of the problem indicated that it was better to strengthen the defence of primary targets to which resources were already allocated. An increase in a nation's defensive strength decreases the enemy's offensive strength. As his strength decreases he has to concentrate on the most valuable targets to get anywhere. These, then, are the ones where defensive effort should be concentrated to win a war.

Methods based on game theory have been extended to so-called metagames where players are unable to place a numerical value on outcomes and can only express preferences between them by way of rank ordering. A further extension to hypergames allows for departures between perception and reality, where players may have false beliefs about the preferences of other players and the options available to themselves and others. This method was applied in a retrospective analysis of the 1940 invasion of France by Germany, where the French command misperceived the Ardennes as a barrier to tank movement and ignored it as a German option. In order to produce a solution this method does require the prior specification of misperceptions. We are seldom aware of these till after the event.

Game theory approaches make several assumptions: that all the possible outcomes can be specified; that each participant is able to assign a measure of preference or utility to all outcomes; that all the variables which determine the payoffs can be specified; and that the value of the payoffs is known. These conditions are unlikely to be met in the field of international conflict. The main contribution of game theory has been to provide a way of grasping the essence of tangled fights. It points to a strategy as a solution. The careful identification of parties to a dispute can clear the mind. The significance of chance in affairs is established with precision. A scoreboard is provided by the notion of the payoff matrix, placing an emphasis on the duality of outcomes. These ideas have helped to clarify complexity and have provided methods to aid choice and decision. The theory has not, however, proved a font of lawlike results for the understanding of international politics.

The state of world affairs and what evidence we have about the minds of statesmen might cause us to hesitate before ascribing consistent and logical ways of thought to them. Theodore White has suggested that Mao was mad from 1960 on. The leader of the most populous nation in the world suffered from Alzheimer's Disease, which brings premature senility. After the fact we frequently find that our leaders' hold on reality is as fragile as that of the rest of us. The nature of the contest for political power, be it a formal procedure, wheeling and dealing or violent combat, encourages those who are arrogant, fanatic or conspiratorial. This possibly makes rulers more susceptible to delusion and folly than the great mass of the ruled — although we do have ample evidence that mass hysteria can be induced and sustained. Game theory assumes a simply defined rationality among players in making decisions or, at least, consistent irrationality. In this it will fail to explain or predict the ways of the world. Neither utilitarian rationality nor altruism can be guaranteed and a large element of chance must be recognised in the ways of the world.

A Spatial Model

For the models of politics considered up to this point, geography is an incidental feature of the world, at most. Often the world is treated as the head of a pin upon which its rulers dance and fight. The introduction of location and space into the specification of a conflict will usually just provide two more dimensions on which an elusive equilibrium or optimum can slip or slide out of reach. If the significance of space in international affairs is to be treated with some rigour, this will be purchased at a cost in the richness of behavioural detail employed. This only gets a geometrical handle on the matter. To deal in general with the lie of the land and the geography of minerals, soil, vegetation and climate is far beyond our analytic capacity. These things can be treated only in a discursive and historical fashion.

The only mathematically stringent spatial model of international relations was constructed by Dacey. His intention was to establish the role of conquest of territory in the evolution of states. To do this, geography was first reduced to a line divided into a number of equal-length regions. For this world, time passes in discrete intervals, during each of which one war occurs between two adjacent countries, being settled by the end of the period. Politics is quite simple. Given a chance, countries engage in war to acquire territory from their neighbours. The winner in any war gets an addition of one region from the adjacent country it just defeated. The events which mark the passage of this process, then, are one-region shifts of the boundaries between countries. Between two points in time, the world changes by one boundary shift, transferring a region from one country to another. Victory brings an expansion by one region into the territory of the loser. Defeat means the loss of one region. For one-region countries this means the end of their existence. Thus, the number of countries may vary from time to time.

The historical process begins with a large number of countries consisting of one region each. War and conquest results in fewer, larger countries of several regions with the consequent disappearance of many countries. War and victory are treated as matters of chance. At any time any pair of countries with a common boundary is presumed to have an equal probability of going to war with each other. For the pair which do so at a particular time, each country has an equal probability of winning. Bigger countries, formed by the combination of more regions earlier, have the same probability of winning as single-region countries. This competitive process continues until the final state of one country incorporating all regions emerges.

The model produces some results which can be compared to the historical record and which test the intuition, forcing us to make conjectures and explore the truth of them. The structure and process of the model were closely paralled by historical work on the coalescence of communities on the narrow Peruvian coastal plain.

In Dacey's model a single country comes into being as a result of a finite number of wars. The number of countries can decrease over time, but can never increase. The average size of country may increase through time, but may never decrease. The rate of fusion of regions decelerates through time according to this formulation. As the number of countries decreases, the probability that the two boundaries of a country will be at consecutive points on the line diminishes. The fewer the number of countries at any time, the fewer the number of one-region countries, the less chance there is of a country disappearing. The passage to the final state of one country progresses rapidly from many small countries to a few large ones and then slows down. There is a prolonged final battle between the last two countries for hegemony. With no other contenders around, they are always at war with each other, with their boundaries going to and fro until one triumphs. These results arise obviously and simply from the structure and assumptions of the model. The source of any departures between the model's predictions and historical cases or instinct can be readily traced to their theoretical origins.

In particular, the assumptions that war is equally likely between each pair of adjacent countries and that the likelihood of victory is the same no matter the size of the nation, could be taken to task. If these assumptions were found wanting in the face of historical evidence, then a case might be made for modifying the model so that the probability of being in and winning a war is related to the size of the country. To do this, however, would be to forgo some of the simple transparency which gives such a model its value. Adding complexity to the picture may soften its outlines so much that no interpretable or refutable theses are generated.

As an apt illustration of the loss of clarity which accompanies the introduction of realistic modifications into a model, consider the matter of fission. One feature which it seems worthwhile to introduce is the prospect of countries splitting up as a result of internal conflict. Empires have fallen. Domestic discord has torn countries apart. The likelihood of civil war and partition can be introduced into the model. The certainty of war occuring in each time interval can be divided between the chance of an international conflict and the chance of a civil war occurring within any country of more than one region. The outcome of a civil war would be the separation of an existing country into two, not necessarily equal, parts, at any of its regional boundaries. In such circumstances it is almost certain that one country would result at some time as the outcome of a finite number of wars and civil wars. The process does not stop here, however: at this juncture there is bound to be a civil war which will divide the one country into two. Eventually, the system will disintegrate into a large number of small countries again. The system will continue to cycle irregularly back and forth between fusion and fission. With such a process it is not possible to make a general statement about the rate of territorial integration which can then be compared with the record of events.

Extending the space of this model to a globe would reduce the promise of producing straightforward, testable results even further, with nothing but the superficial trappings of geographical reality to be gained. It does seem that in order to theorise about the geography of international conflict, a less general, deterministic representation, circumscribed by a set of particular and elaborate assumptions about the behaviour of statesmen, is the most we can aspire to for now. The hope of producing unequivocal, testable predictions about the geography of politics is slight. The best we can aim for is to construct a simple mechanism to represent the world. This may yield some insight as an instrument for telling the parables of comparative statics.

To suit the circumstances of our time, our attention should focus perhaps on the final stage of duopolistic competition in Dacey's basic model and the fission of large empires which occurs when he allows for internal strife.

Readings

The economic synthesis of the realist and welfare schools of thought is done in:

Gilpin, R. *War and Change in World Politics* (Cambridge University Press, Cambridge, 1981)

The history of game theory is dealt with in:

Dantzig, G.B. *Linear Programming and Extensions* (Princeton University Press, Princeton, NJ, 1963), Chapter 2

The mathematics of game theory are treated in:

Hadley, E. *Linear Programming* (Addison-Wesley, Reading, Mass., 1962), Chapters 11 and 12
Intriligator, M.D. *Mathematical Optimization and Economic Theory* (Prentice Hall, Englewood Cliffs, NJ, 1976), Chapter 6

The application of game theory to military problems is examined in:

Quade, E.S. (ed.) *Analysis for Military Decisions: The RAND Lectures on System Analysis* (North Holland, Amsterdam, 1970)

Metagames and hypergames are defined and applied in:

Bennett, P.G. 'Toward a Theory of Hyper Games', *Omega,* vol.5., no.6 (1977), pp.749–51
Fraser, N.M., and K.W. Hipel 'Solving Complex Conflicts', *IEEE Transactions on Systems, Man and Cybernetics,* vol. SMC-9, no.12 (1979), pp.805–16
Howard, N. *Paradoxes of Rationality, Theory of Metagames and Political Behaviour* (MIT Press, Cambridge, Mass., 1971)
Isard, W., and C. Smith *Conflict Analysis and Practical Conflict Management Procedures: An Introduction to Peace Science* (Ballinger, Cambridge, Mass., 1982)

The spatial model discussed is from:

Dacey, M.F. 'A Model of Political Integration and its Use in the Reconstruction of a Historical Situation', in K.R. Cox, D.R. Reynolds and S. Rokan (eds), *Locational Approaches to Power and Conflict* (Sage, New York, 1974), pp.213–30

5 DISTANCE AND POWER

A Simple Model

We can perhaps learn something from putting relations, structures and processes which we remark from casual observation into a formal language. Most importantly, it brings them under closer scrutiny in a setting where rules of evidence and proof apply stringently. If nothing else we may see what is missing in our understanding or what circumstances and causes of events cannot be captured in simple formality. There is always a danger, of course, that a simplified likeness will be taken for reality itself and that such efforts at theorising will be taken too seriously. To hope for predictions of events comparable with the precision of physics in matters which arise from sums and products of human choices is foolish. Nevertheless, if we do not set up and argue a strong case our knowledge is not likely to advance.

The first need in building a model of the state of affairs is to point up a clear driving force for those who play a part in the plot. In this vein, our world will consist of some aggressive people in charge of some countries' fortunes, who seek to control others by spreading their nation's power in the form of military might, diplomatic or administrative presence, economic penetration and propaganda over the territory of other passive states. Let us suppose that in each expansive country there is one person with the power to act in foreign affairs, and that they all act so as to dominate as much territory and as many people as possible. To keep the geography simple, we can assume there is a uniform density of people over an undifferentiated world, so that territory and people are the same thing, apart from some densely occupied core areas which can be treated as points on the surface. The object of policy is, then, similar to the object of getting as big a share of the market for your product as possible, which is sometimes attributed in economics to decision-makers in industries with a few big firms resonsible for all of the output.

In competition with other powers, territory is dominated by having more power in a particular area than anyone else. Leaving questions of the constitution of dominion and the whereabouts of national boundaries aside for now, we can consider the flow of power from the cores and the boundaries of spheres of influence in simple, geometric terms.

The next question to arise concerns the nature of power. What is the nature of this force which flows out from the cores? Power may take the form of military might, administrative and police control, diplomatic influence, flows of money and goods, bribes or threats or ideas spread over the territory of other countries. Whatever its components, we can distinguish two varieties of power exerted by the time it takes to apply them. For our treatment of these matters we need then to lay out a time frame. It is convenient to divide time into three states: the short term, the mid-term and long run. The long run is that stretch of time in which many of the things held constant for short- and mid-term analysis are allowed to vary. For the short run we can assume that the total amount of resources which can be used for the generation of power in any country is constant. In both the short and middle run, the technology for the spreading of power and the total resources of the nation are unchangeable. In the middle run, the amount allocated from the fixed total of resources and its distribution between various uses of power may be changed.

In the short run, the response to a challenge or opportunity to move into some piece of territory is met by sending out a force, or goods, or propaganda, or a mission. For this short-term use of power the frictional cost of overcoming distance is the determinant of outcomes. The cost of holding territory is a given for this time span and we might call the energy on which choices are made in this period 'missionary power'.

In the middle run, influence over a piece of territory is fixed by building up networks of influence or control. The cost of covering territory with a mesh for communication and movement and working this is the most important variable for this period of choice, rather than distance from the power core. Choices are made of how much of the resources of the nation to allocate to exerting power with a given technology and how to divide these between missionary power and what we might call 'territorial power'.

In the long run, the methods used to generate both missionary and territorial power may change and affect the competitive positions of different nations, as indeed may the total resource bases of nations as well as the amount put to spreading the influence or control.

In the short run, there is a given amount of energy which may be used for missionary work. The choice is where and how much to exert this force. The amount of resources deployed to command territory is fixed and not a matter for choice. In the middle run of events, it may be possible to change the amount of resources given to spreading power and to change the balance between missionary and territorial uses. The main matter for decision, however, is the extent of networks of control. In the long run, the choices to be made include not only those between arts of employing power, but also those involving paths of economic growth and levels of zeal for mastery of the world, between glory and bourgeoisdom as well as between guns and butter.

The Short Run

Considering the short run first, let us see what we can expect to happen when any one core state with an urge to swell its domain is presented with opportunities to make its presence felt in a variety of places. If the total of its missionary power resource is not sufficient to exploit all of these, choices must be made. If a leader has some given quantity of resources to apply to transporting and exerting influence then, clearly, the potency of power delivered will diminish with distance from the core as more energy is consumed in transport. Even propaganda must pay the cost of distance friction. Beaming 14 hours' propaganda a day at Cuba from Marathon on the Florida Keys, Radio Marti will cost US taxpayers over 10 million dollars a year for the first two years of operation. The cost of distance has two parts, a terminal cost and a line-haul element. The latter is a function of distance, while the first has to be undertaken no matter how far or near you travel. The net effect is that transport or transmission cost functions can be drawn with a constant intercept and a positive inclination with distance as in Figure 5.1.

Figure 5.1: Cost of Transmitting Power

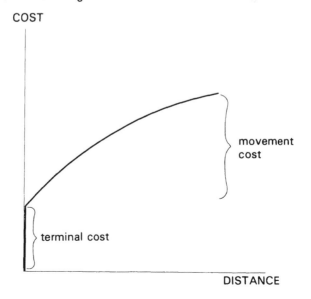

Since there are often economies to be had from spreading over a greater area some fixed components of costs which are not related to terminals, such as time-rate labour or vehicle costs, transport costs usually have a distance taper so that they would be convex upward, as in Figure 5.1. If this is a reasonable picture, then the potency of power delivered by some fixed total amount of energy is likely to vary in an inverse fashion to this, as in Figure 5.2.

Figure 5.2: Power Available

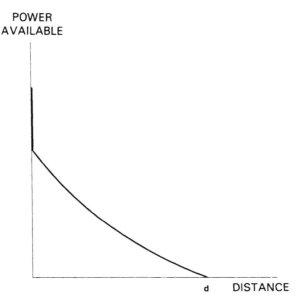

Eventually, distance costs may eat up all of the available reserve of energy and the quantity of power left to be deployed will be zero at some distance, d, from the core. If two opportunities are at different distances from the core then, obviously, power may be brought to bear more cheaply at the closer rather than the farther one. If two core areas are in competition over the opportunities available and use the same amount of energy available, then the reach of the respective powers will bisect the array of opportunities in territory lying between them. If the opportunities were arrayed along a line between core A and core B, their respective spheres of influence would be as in Figure 5.3. We can presume that the outcome of competition in the range d′d where either A or B could deploy power, is determined by whoever has the more power to exert at a particular point. Thus, from d′ to a, A's power is greater and it prevails, while from a to d, B is the winner. The boundary between their spheres of influence, a, is found where they match each other in power.

Figure 5.3: Spheres of Influence

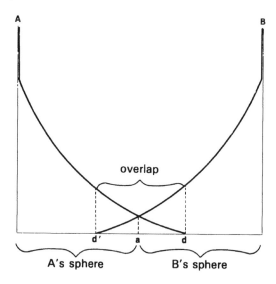

If A has a greater total amount of resources to apply but still uses the same means as B to deliver them, its reach will be more extensive, as shown in Figure 5.4. Even if the slope of the two delivery curves remains the same, the higher intercept value of A, representing a greater amount of energy available, will result in a boundary line, a, which is closer to B than A.

Figure 5.4: Unequal Resources

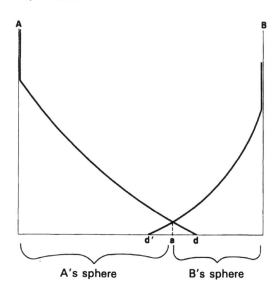

If, however, B is a sea power and has lower costs of delivery than A, a land power, the situation may be reversed, even if A has a greater resource base. In Figure 5.5, the flatter delivery cost curve of B offsets the greater energy available to A and results in a boundary closer to A. Obviously, this result would hold only for opportunities equally accessible by land and sea. It would be possible to extend the model from competition for points on a line to a two-dimensional surface and to introduce the configuration of land and sea into the problem. It would be feasible to give each core a combination of land and sea transport capacity or to apply different transport cost rates for movement over land and sea, allowing for the costs of transfer from one to the other. For our purposes, however, this would be a needless complication.

Figure 5.5: Unequal Delivery Costs

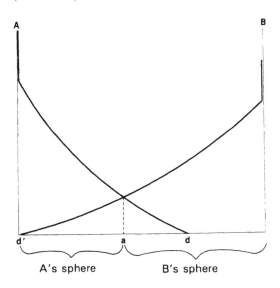

The Middle Run

Decisions to extend the reach of a power core into a new territory may be made in terms of short-run costs at the outset, but the more lasting outcome will reflect the ability of power cores to maintain their influence over territory. If the spreading of hegemony depends upon delivery costs, the maintenance of it depends on the local costs of territorial control. The choice of how much to invest in networks of control has been made a matter for the middle of run time. How far to extend the mesh of a core is a choice made over the same time span as the decision on how much to invest in total in the thrusting, missionary power to exert influence. In the same time frame the balance between these two forms of power is struck and the total of resources to apply to these activities rather than other uses is resolved. The choice between guns and butter and the selection of varieties of gun are both made in the middle term.

If for the present we retain the simple objective of gaining influence over as much territory as possible as the purpose of those who govern, we can explore the significance of distance to the exercise of power by following the extension of a network of influence outwards from a core. If action in the short run establishes a presence in a territory, the next step is to consolidate it. Let us suppose for now that the total amount of resources to employ to satisfy global ambition is determined by a political process in which the outcome is a matter of the internal character of the core nation. To simplify matters further, let us suppose that the amount of resources to be used in the form of control power over the middle term is established on the basis of some expectations of the rulers. This given amount of resources must be used to establish channels of control, exchange or information and operate them. If action in the short run marks out a sphere of influence, the next step is to consolidate a presence on territory within its bounds. If the resource base and technology of the core remains constant in the middle run of time, to extend its net of control more widely would spread the same amount of territorial power more thinly on the ground. Now we must provide a geographical frame for our model. To keep things clear, we presume that the population and surface to be invested is the same everywhere. The world is a uniform plain with a constant density of people, except in the intense clusters of power cores. These we take to be so concentrated as to be points in the plain. It is reasonable to assume on this basis that to get the same response from everywhere it is necessary to deploy a similar amount of power everywhere within a desired radius of influence. If we take a fixed amount of territorial power and apply it uniformly to a wider and wider radius of land, then the quantity of power per unit area or per person will diminish as the inverse of the square of the radius, as in Figure 5.6.

If we label the total quantity of power P, the density of power will be P/T where T is the area within a given circumference from the core. T is given by πr^2 for a circular region of radius r. Density of power will thus be $P/\pi r^2$. As r increases, the density of power will diminish from infinity at zero radius along a concave curve, so that it will approach zero only as circles are drawn more and more widely. It would

Figure 5.6: Diffusion of Power

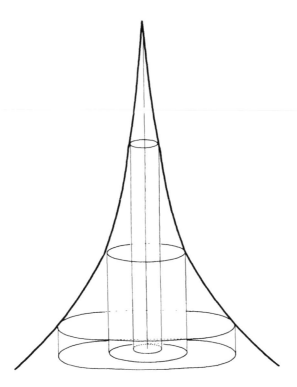

be possible to introduce geographical differences in population density and terrain into the picture at this point but, again, it would confuse things. It would also introduce a superficial reality which might lead us to credit the model with more representational power than it warrants.

If there is a limit of effective density of power, below which control or influence is lost, then clearly the potential sphere of influence is curtailed within a finite horizon, as in Figure 5.7.

If two cores are spreading their meshes of influence towards each other and the effective lower limit is hit, even if their missionary power led to one boundary between them, their territorial power potentials may not overlap and there will be two boundaries leaving a neutral belt between the spheres of influence, as in Figure 5.8(a). If two cores of similar power do not meet such a limit before their potential fields of influence overlap, then, as with missionary power, conflict may ensue over the range of overlap and the most likely outcome would be a stalemate in the vicinity where their power is equal, which would be a straight line, as in Figure 5.8(b). If one core has more power than the other, say A's is greater than B's, the boundary will be curved around B, as in

Figure 5.7: A Power Threshold

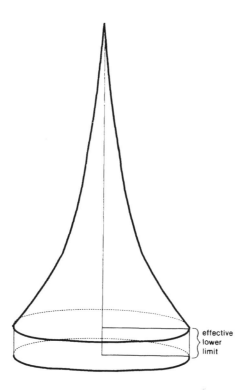

Figure 5.8(c). If B has a more efficient technique for deploying power, then this may be represented by a flatter decline of potence with radius which we may think of as a constant k applied to P to give $Pk/\pi r^2$. This may be enough to reverse the concavity of the boundary of influence, as in Figure 5.8(d).

In the middle term, we can consider motives which are more elaborate than the headlong drive to dominate as much of the world as possible. This is possibly too raw. Over this time span the total of resources to allocate to missionary and territorial power is decided upon. We could credit leaders of nations with a more calculating, hedonistic attitude. The accounting then comes as a balance of costs and yield. The yield may be in the form of rent from land, taxes or profits from trade. Decisions are made so as to maximise the net yield of empire or a sphere of influence. Whether or not actual choices have been made in such an economically sophisticated fashion, economic gain has been held up as the goal of imperialism or hegemonism. We can ask what might happen if this was pursued in a rationally calculated fashion. With such a motive there may be a radius of influence which is best at less than the limits

Figure 5.8: Competitive Variations

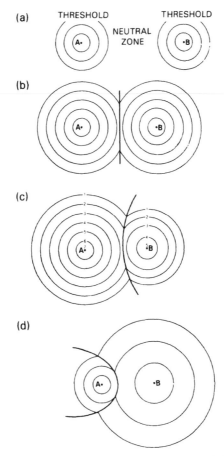

of a boundary established by competition for global dominance.

We can keep the assumption of a uniformly dense, even plain and consider the total power applied and the total yield of territory as they vary with distance from a core. This makes it evident that at the outset of the expansion of a sphere of interest, costs will not be covered. The yield of wealth from territory starts from zero at the origin and increases in proportion to area. There are some start-up costs involved in establishing an empire or sphere of influence, the terminal costs of missionary power and, possibly, some overhead administrative costs of territorial control. These will remain unrequited over some radius from the core. This is not only an artificial product of the representation of affairs by the smooth geometry of curves on a graph and the departure this involves from the discreteness of national frontiers and the blocks in which territory is usually acquired or penetrated. The fact remains

Figure 5.9: Yield and Expenditure

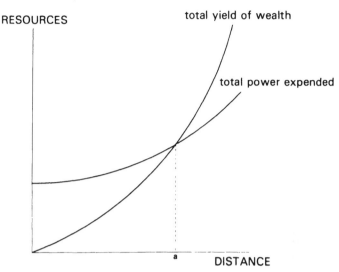

that an ambitious cadre would have to amass the means of conquest or insinuation before it could acquire any territory. If it was economically calculating, it would have to have a reasonable expectation that the yield of wealth would exceed the outlay of power at a modest radius from its nucleus before it started. In Figure 5.9 this point is reached at radius a, beyond that radius the total yield of wealth exceeds the total outlay of power. If this were not to happen, there would be no economic rationale for domineering tendencies.

If the curve of yield does cut the power from below in this fashion, in the geometry in which we have cast things it cannot cut it again. Our asumptions are such that these are both monotonically increasing functions of distance from the origin, i.e. their values will continue to increase as distance increases. If this is so, the temptation will be to continue to thrust outwards, for the increase in yield will always be greater than the increase in power spent. There will be no limit to desirable territorial extent in terms of return over outlay diminishing at some point. The only limit must be the competitive one of conflicting in power with some other aggressive cadre. On a sphere of uniform density the functions describing yield and expenditures could cease to grow exponentially at a radius which extended the sphere of influence to a hemisphere. They would grow at a decreasing rate thereafter until the entire sphere was covered, at which juncture they would have flattened out.

There may be an economic limit to the extent of a sphere of influence with regard to missionary costs which do vary in incremental fashion

Figure 5.10: Friction and the Limits of Empire

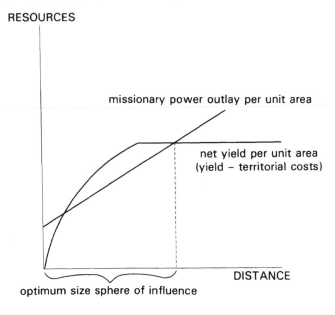

optimum size sphere of influence

with distance from the core. This is contrary to our assumption about territorial costs, which we deemed to be constant everywhere. If the missionary component of the cost of influence is very high, it may be that at some radius the resource expenditure per unit area will exceed the yield of wealth per unit area which is a constant. This possiblity is shown in Figure 5.10.

There may also be a limit in terms of the workings of the government in the core area. These may suffer from the political and administrative equivalent of scale diseconomies in production. If this were so, the efficiency of the working of government could be represented by a U-shaped curve. When the unit resource cost of the service of government is plotted against the amount of government, which increases as the territorial extent of the state increases, there is a decrease at first. As the functions of the state expand, there is a decrease in costs as scale economies reduce the burden of additional territory. This continues down to some lowest level of resource costs beyond which additional units of territory to govern cause increasing central government costs. This would represent the effect of congestion on a fixed set of insititutions and procedures, perhaps, and their technical and political limitations. If this is so, there may be a point where the cost of government, or influence per unit area, begins to exceed the constant yield of wealth per unit area we have postulated for territory, as in Figure 5.11. Whether

Figure 5.11: Diminishing Returns to Empire

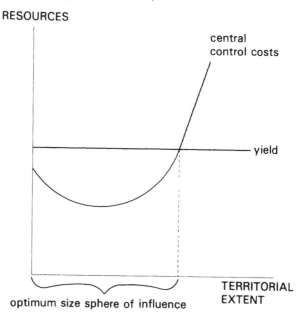

or not rulers take actions advised by calculations of yield and cost, their budgetary process may constrain them in this direction.

One other choice we left to be made in the middle span of time was the balance between missionary and territorial power for a state. The decision to switch resources to reaching out rather than to consolidation may enable one power to gain a short-run advantage in prospecting for dominion, which we could represent as a shift in the boundary of the balance of power from a to b in Figure 5.12 as B extends its missionary power. This extension must be bought at a cost of reducing the ability to control territory or some other function of the polity or by using more resources for imperial purposes. This may have a deleterious effect on productivity and social welfare. These can rebound in stirring political unrest in the core area. Khrushchev's adventures overseas after 1956 and the build-up of the blue-water navy by the USSR might be viewed as a shift of this kind, and the costs involved are easy to see. There are historic examples of a shift in the other direction, concentrating on intense command over a smaller territory and pulling in the tentacles of a longer reach of involvement. The Byzantine empire is an example of this transition. But here we merge into the long run, for the contraction of Constantinople's reach was accompanied by technical and social changes. It involved the evolution of a territorially dispersed, military

Figure 5.12: Extending Reach

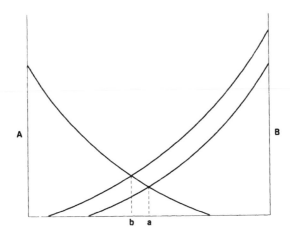

aristocracy using big horses in heavy cavalry formations to defend the empire against the light horsemen of the steppes.

The Long Run

In the long run, the quantity of power available in a core region is a variable, as are the means and arts of war, commerce and persuasion. The relevant choices for the long run involve possible paths of economic growth and amount of investment in technical advance.

If a power core produces a technical improvement of the means of delivery of missionary power, as with the development of efficient cannons on shipboard by European sailors at the close of the fifteenth century, its potential boundary of power will push that of an unimproved rival back. Suppose that in Figure 5.13 A improves its delivery of power while B does not, then A's gradient of power is reduced and its reach pushes the sphere of B back from a to b. We might view the Prussian general staff's exploitation of railways for military purposes from 1850 on, which gave them a distinct edge in extending their hegemony over Germany and dominating Europe, in this light.

The efficiency with which territorial control is exerted may be improved, or may deteriorate, multiplying or dividing the effectiveness of a given amount of power by raising or lowering the constant k in $Pk/\pi r^2$. The reform of public schools in England, led by Dr Arnold of Rugby, introduced geography, modern history and languages along with

Figure 5.13: Improved Delivery

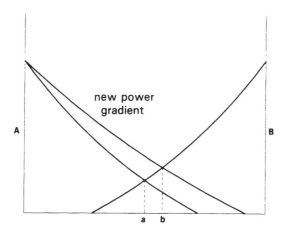

team sports into the curriculum, and so turned out an effective class of administrators and district commissioners to run an empire on the ground. In reverse, the abolition of 1638 of the janissary system in the Balkans, whereby the Turks had recruited Christian lads to their army and administration, began the rot of the Ottoman empire. The gap between Turkish landlords and Christian peasants widened as the bridge of officials who had spent their first 12 years in Christian villages was cut. This also created a stream of unemployed, ardent youths, who headed for the hills and turned to banditry.

In dealing with all kinds of longer-run changes, we are getting far beyond the tenure of individual rulers and are dealing with ill-defined corporations. The labels 'elite', 'establishment', 'ruling caste', spring to mind. No matter the legal process for the transmission of authority through time, there is strong evidence of the continuity of a leading group at the head of most societies. This may be subject to drastic change or total replacement. In other instances, a certain set of ideas has remained entrenched as power is retained within a closed circle. Elsewhere, potential disruption is avoided by co-opting dissidents and giving way on some issues. How, when and why these groups exist, persist or change is the marrow of history. The make-up and workings of such collective responsibilities are the subject of conjecture, but remain unknowns. Whatever we call the groups who make choices in the long run, we must keep in mind that the people involved, their desires and views of the world may change radically over this time span. This means that there are no givens and any statement must be hedged about with the assumption that all else is equal. If the short and middle terms are defined by assumptions of constancy, the long run is by definition the time over which everything could possibly change.

An Arms Race

An arms race between two nations provides the setting for following the transition from short- to middle- to long-term choices and actions. We start in Figure 5.14 with two nations, A and B, in a balance of power. The magnitudes of their power, A' and B', are similar and their spheres have a boundary, a. To explore the reactions involved, we introduce a shock to the arrangement in the form of increase of the power of A to A''. With greater means of influencing other nations, including arms, the boundary between the spheres of influence of A and B is shifted to b. In the short run, the rulers of B may be able to counteract the thrusts of A's forces and agents in some places by completely expending their reserves, but inevitably the dividing line is pushed back to b. In the middle run of events, B's leaders may respond by increasing their expenditures on the means of influence, especially on trained men and weapons. This response might possibly more than counter A's initial advance, thrusting the boundary back to c. In the long run, an exchange of these thrusting efforts may encourage the elite of A to devote research and inventiveness to developing a new means of delivering its power and expanding its budget for doing so. Historically, this has usually taken the form of new tools and vessels of war. This investment will rotate A's gradient of power upwards, pushing out its influential frontier with B to d. It has seldom

Figure 5.14: An Arms Race

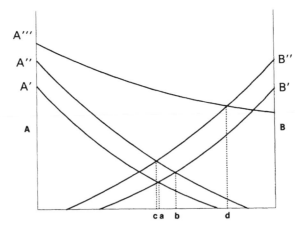

taken long for a nation in B's position to acquire or duplicate A's advances and surpass it, setting up an upward-spiralling sequence of counteractions. So far in man's history, such races have usually come to an end in the use of the weaponry developed in all-out war. This is a chancy and destructive means of resolving conflicts which does not necessarily yield a net gain to the victor. With the weapons available to the two chief contenders for global leadership at present, the outcome of such a resolution could well be the death of a large part of mankind and the destruction of the life-supporting capacity of much of the earth. For whoever was left, victory would not have much yield nor, indeed, any meaning.

Location and War

Most of the conflicts of the last 30 years have arisen in the crush zone between the great powers. The force fields of the hegemonies may be thought of as extending out from their cores, overwhelming smaller nations with their power, surrounding the spheres of influence of lesser powers and lapping against each other at the edges. In this picture, the rulers of each state dominate the territory over which its potential to exert force is greatest; for the great powers this domain may extend far beyond their formal territorial limits. This image of the world, however, contains no impulse for violent conflict. Rather, it conveys a sense of stability and rational discretion, with the balance of power set along the lines of intersection of force fields.

Figure 5.15: Uncertainty and the War Zone

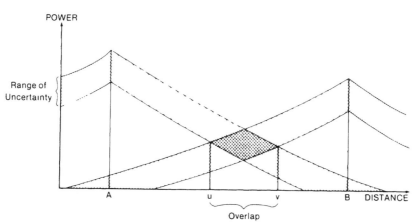

The reality of war is spawned by doubt. An aggressor sees a chance of success where there is some doubt about the equality of power. On the other side of the equation, the defender does not know for certain that he will be beaten and makes a fight of it. Such uncertainty can be introduced into the schema offered here by construing the gradient of power not as a line, but as a strip of some width, as in Figure 5.15. At any distance from its core, a nation's power potential is known to lie between certain limits, but its exact magnitude could be anywhere in that range. This produces an overlap of spheres of influence, between u and v in the figure. Over this interval there is doubt about which power could exert the greater force. This uncertainty, especially at the outer fringes of the fields of dominance of great powers, creates the temptation and opportunity for local enterprise in war. The war between Iran and Iraq in the Shatt el Arab, the battles in the Lebanon, the fight between Ethiopia and Somalia over the Ogaden, the clash of Moroccans and Polisarios in the Western Sahara and the Argentinians and British in the South Atlantic could be all characterised in this fashion.

Even outside the area of overlapping great power potential, uncertainty provides the occasion for violent conflict. Dissidents may seek to overthrow the incumbent government in parts of the world clearly dominated by one great power because it is not certain that their ability to achieve their ends with force is less than that of the government or the dominating power. We might view the cases of Guatemala, El Salvador, Nicaragua and Afghanistan in this light. Despite the location of these struggles well beyond his direct military reach, the distant hegemonist is tempted to get involved. Here is an opportunity to discomfit your chief competitor in his own backyard, which carries little risk as long as the prospect of direct confrontation is avoided. Thus, such insurgents often win the encouragement and clandestine aid of a great power on the other side of the world. The government of the USA accuses the USSR of such exploitation in Central America, while the Soviet government makes similar accusations about US involvement in Afghanistan.

Shortcomings of the Model

The state of affairs in which we live has collapsed the intervals between short-, middle- and long-term actions in a frightening fashion, so that the long-run response time in technical terms in a matter of years, not decades. The USA had multiple-warheaded missiles ready for action in 1973. The USSR tested them first in 1973 and had them sited by 1975. The treatment of time in discrete intervals is subject to a tight compaction when we deal with the nuclear competition. The length of the intervals is not an historical constant.

What has been laid out here is far from a well-jointed theory. It is merely the use of a simple mechanical framework to organise some thoughts on the competition which overshadows all of our lives. To tidy it up into more elaborate formality and tie up its loose ends would give a spurious appearance of logical tightness and rigour, carrying the false hope of simple solution. The competitive process has been left vague and boundaries are allowed to arise from the equality of deployable power at some places. The final recourse, war, tends to be decided more drastically one way or another, and the expenditure of resources involved goes to limits beyond any calculated balance of costs and benefits of likely outcomes. The mad logic of war pushes long-run material advantage aside in the immediate fever of seeking victory. Efficiency is defined in terms of the need to win.

To have nuclear missiles which can strike anywhere on earth in a matter of minutes at their command, gives the leaders of the major powers the last word and constitutes influence in a form which seems to deny the significance of distance in diminishing power. Of course, even with these weapons there are limits to their capabilities which make the location of launch sites and the selection of targets a matter for strategy. Beyond that, the stalemate which the destructive prospects of these weapons has brought about, creates an uneasy ring within which the hegemonists compete in indirect fashion for the attention or sympathy or resources of the territory between them, still subject to the wear of moving over the surface of the earth, through its waters or atmosphere.

One matter which the geometry we have employed so far in this chapter does not account for well is the extent and geography of core areas and the deeply entrenched reality of political borders. Political territory comes in blocks. It might be that a discrete treatment of political space, as intervals of a line or squares of a grid, is more appropriate. This would really produce only superficially more realistic, finite outcomes from the conflict of overlapping power gradients; it would not add any greater depth of understanding.

The Euclidean distance used in this geometric representation of affairs does introduce specious simplicity into the analysis. Distance might be measured more pertinently along routeways than great circles. The relevant distance should reflect the lie of the land and ocean and the different friction offered to movement by uplands, lowlands and the sea. The cost of overcoming distance is not a constant, even in the same medium. The marginal cost of transport diminishes as journeys lengthen. The question of which points to measure distances between arises. Are the closest border crossings, major cities or military assembly points of nations the relevant terminals? Or are the seats of government most appropriate? Obviously, the relative positions of nations with respect to the great continental powers change depending on which end-points and metric you choose in measuring distance. What is close to Vladivostok is not close to Moscow. Hanoi is 2,000 miles from the closest point on the Soviet border, but over 5,000 miles from Moscow as the crow flies along the length of the Himalayas. San Diego is 9,000 miles away, but over the ocean. Since the cost of carrying material on the sea is about one-tenth the cost of road haulage, however, the USA is closer to Vietnam than the USSR, in surface transport cost terms.

Another assumption implied by the drawing of power gradients is that all space is equally attractive and that nations project power outwards symmetrically. In reality, some pieces of the earth's surface are much more desirable than others and the might of empires has been concentrated on holding these rather than everywhere within a similar radius. The oilfields around the Persian Gulf generate a major departure from symmetry as the government of the USA strives to exclude the influence of the USSR from the region, even though the USA is twenty times further away than the nearest point on the Soviet border.

The most unfortunate shortcoming of treatment of the world in simple continuous terms is its overemphasis on the importance of the major powers in world affairs. Once a nation has fallen beneath the power envelope, it loses identity.

The World as a Network

Pictures of gradients of power do contort our vision and detract from our understanding of how nations interact. Continuous fields of influence concentrate attention on the notion of domination and, thus, the needs and reactions of the dominators. The reality of our dealings with one another on the face of the earth, be these political, social or economic, is better represented by vectors rather than waves. The number of people and the connections between them are finite. We are countable and the links among us along which something is passed back and forth are limited in number. If we were to construct a table with a row and a column for every living person, with an entry in the cell formed by the intersection of the appropriate row and column for any interaction of any kind between people, then the vast majority of cells would be empty. Although radio and television broadcast events and selected information and views around the world with a lag of only seconds, in terms of replying to signals with signals, there are a limited number of channels of communication. Even telephone ownership is not nearly universal.

The realm of interacting human intelligence has been characterised as a continuous field. Mackinder called it the 'psychosphere' in 1937 and Teilhard de Chardin called it the 'noosphere' in 1955. This realm does, however, operate more nearly like a network with a finite number of start- and finish-points and links. Especially at the international level of interaction, there is a scanty mesh of connections joining the few people with the power to influence and decide for others. Interaction over this network takes place both directly between co-operating or competing parties and also indirectly through go-betweens, proxies or clients. The transmission of power and prestige varies in strength according to the character of the sender and receiver, the means used to channel the flow and the nature of any intermediaries used.

To think of the world as a network of relations makes it easy to depict power as resident in and arising from many more than the few, big, aggressive nations that dominate world politics today. The knots in the net of power need not only stand for national governments, but could represent a variety of places where people have power to dispense. Not only would it be possible to point out links between geographically specified sources of power in different countries but, also, links between regional sources of power in the same country could be taken into account. Poles of power outside the authority of the state — churches, factions, businesses and organisations which operate across national lines — could also be brought into the picture.

If it were possible to identify the significant links in the world and quantify the action and response relations which take place along them, then there would be a way to trace and measure the effects of a disturbance starting at any place in the system. To do this completely enough to predict events with precision will always be beyond our capacity. Power is not a well-defined, homogenous substance. We do not know where it lies nor how it is exerted nor the equations which govern the resolution of power conflicts. Power is a shorthand term for a multiplicity of ways of exerting violence and influence or of stimulating sympathy. Each of these forces has its own network for transmission. The fact that the final resolution of the competition between the powerful has most usually been war, a drastic and near-run thing, offers only the prospect of a guess at chances of outcomes. The value of the network picture of the world lies not in its promise as a means of forecasting events, but in its portrayal of a more complicated world than the crude model of the first part of this chapter, which translates the language of great power politics into geometry. Amongst other things, the idea of a complex system holds out the hope of a world in which local jolts which disturb the prospects of survival will be absorbed with little harm by dissipation through the network.

To think of the world in these terms offers no chance of neat, comparative static analyses of the likely outcomes of disturbances to the system. The power gradient formulation lends itself to this use. The network model inspires us to look for complexity and to seek to quantify expressions of the relationships involved. In this alone it may be more useful, forcing us to look at the facts of history and geography which provide the setting for political events. This provides a counter to the notion of a headlong destructive contest between two big countries, crushing all that lies between them. This is conjured up by the idea of continuous fields of influence spreading from major poles of power. For statesmen to think in terms of spheres of interest which override national boundaries seems likely to generate Pavlovian responses to any departures from the status quo within each hegemonist's definition of his turf. In essence, the power gradient view deals in dominance relations. The pole with the highest power surface at any point in space dominates that part of the globe. It is possible to start with a number of poles in mind, but their influence rapidly gets lost beneath the enveloping spheres of the major powers, as in Figure 5.16(a). This figure presents the picture of power gradient competition along a line, with A and B, the dominating poles, accounting for most of the upper envelope of power, but with intervening dependents, such as C and D, and interrupting autonomies, like E and F. Projecting this into two dimensions produces a map in Figure 5.16(b), where the extent of the dominated fields is even more pronounced.

A network picture of the setting of foreign policy decisions is suggested in Figure 5.17. This may encourage treating each cause of disquiet on its merits in terms of an explicit, rational notion of national interest. In this diagram some of the possible connections between four nations are detailed. Nations A and B have both direct connections between central governments and regionally specified interest groups and indirect connections via other governments, interests and multinational organisations. Some smaller nations, such as D, only connect through their dominator or the likes of the UN. Others, such as C, have direct connections with both great links. In reality we are faced with several networks with different friction factors and structures.

Figure 5.16: Force Fields

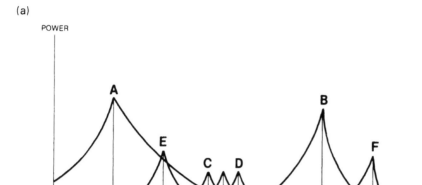

The complexity implied by this image of the world is not easy to con-
vey in words which will win votes or emotional support. We conven-
tionally divide the world up into 'them' and 'us' and reckon that who
is not with us is agin us. National and international politics are treated
as a game of football. Institutions have been invented and fostered to
sustain duality. This may be a reasonable simplification for national
politics. It is a dangerously deceptive one for international affairs.

Figure 5.17: A Network

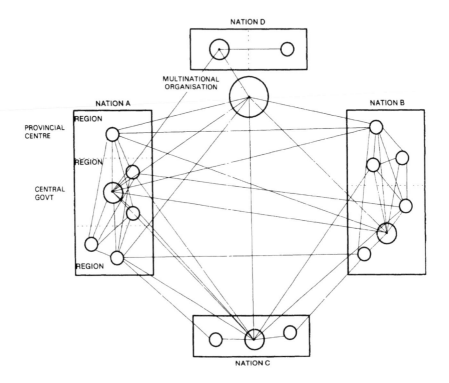

6 MEASUREMENT

Power

To test the truth of different views of world politics it is necessary to measure and map the geography of power. The generalisations on power which have been made can be interpreted for the most part as propositions about the extent and shape of the fields of influence of leaders of nations. Power is that in a person or group of people which enables them to act on others. The word is an abstraction for the variety of means with which we can control the behaviour of our fellows. There is no homogenous flow of political power which can be metered and recorded. No uniform packages of matter or waves of particles register its force. The reality of power consists of flows and waves of many kinds moving between people in different places. Power is exerted through armed might, diplomacy, the manipulation of economic relationships, the collection and use of information and the spreading of ideas. To a large extent power consists of images established in other people's minds. Our attitudes and actions are governed by perceived qualities of people and the world. Bringing about perceptions in other people to induce suitable attitudes and behaviour among them, is the most potent exercise of power. This can be tried with demonstrations of violence, as the USSR is doing with little success in Afghanistan and the USA did more successfully in Grenada. Threats of violence, veiled or otherwise, can be used, as the USSR did in the case of Poland in 1981 and the USA is doing in the case of Nicaragua now. Shows of benevolence or bribery, such as the USSR is giving in Iran and the USA is making with its Caribbean initiative, may do the trick. Witholding economic favour by embargoes on trade and investment has been used to demonstrate disapproval of bad behaviour and to encourage good manners. In November 1983 the US foreign aid bill forbade giving aid to any country that the president finds 'is engaged in a consistent pattern of opposition to the foreign policy of the United States', as measured by its UN voting record. And all the time high ideals and altruistic sentiments are invoked and sounded to win the hearts and minds of all mankind while the opposition is anathemised as evil or misguided. All of these means are instruments of power. Their concrete expressions vary in quality as well as quantity. The transactions involved are sometimes secret and not publicly visible at the time they are completed. Some exercises of power may be measured as flowing volumes of words, paper, money, goods, firepower or soldiers. But volume alone is no guarantee of getting the desired response. It is impossible to know with any precision the effect on the people targeted. The amount of fear, greed or spiritual uplift generated or gratified can only be seen in the doings and sayings of the receivers and the agreement of their words and action with the will of the sender of power. Even then, people are capable of simulating conversion to survive or for material gain, only to revert to their old ways when the pressure is off.

In the second half of this century, conquest and the explicit assumption of imperial authority over territory has been ruled out by the appalling nuclear stand-off between the USA and USSR, the greater appeal to popular support in politics and by institutions of international law. The chief means of spreading and exercising power, then, are indirect and appeal to the sympathies of other rulers and peoples. Strength cannot be measured by the territory and population under one flag or another. Yet power is exerted from outside over the populations of nominally sovereign, independent states. To express this reality we can only think in terms of recording the geography of power with subjective maps of the affinity of the governments or people of different nations to a superior power.

The membership of nations on one side or the other, or their neutrality, in the competition for world leadership can be mapped for any moment of time. The current of informed opinion, government actions and pronouncements can be used to place nations in the rival camps or leave them neutral, as in Figure 1.1. Any such threefold classification of countries is bound to arouse disputes and is subject to daily changes. Drawing a three-colour map of the hegemonies of the USA and USSR and the non-aligned nations between these is likely to feed rulers' delusions of grandeur and the anxieties that accompany these. A longer-term reality might be captured by allowing for shades of pro-Soviet or pro-American leaning, as well as for neutrals, as in Figure 6.1. To do this with any precision it would need many colours to express the subtleties and vagaries of the positions which national leaders take and public opinion supports. Whatever the details of uncertainty and fluctuations in the colours of areas through time, what such a map of the response to power indicates is that influence does in general diminish with distance from the power cores. There is not a random scatter of affiliation. Nations whose leaders are of similar persuasion do tend to form geographical blocks. There is a less than perfect correlation between affiliation and

Figure 6.1: Degrees of Influence

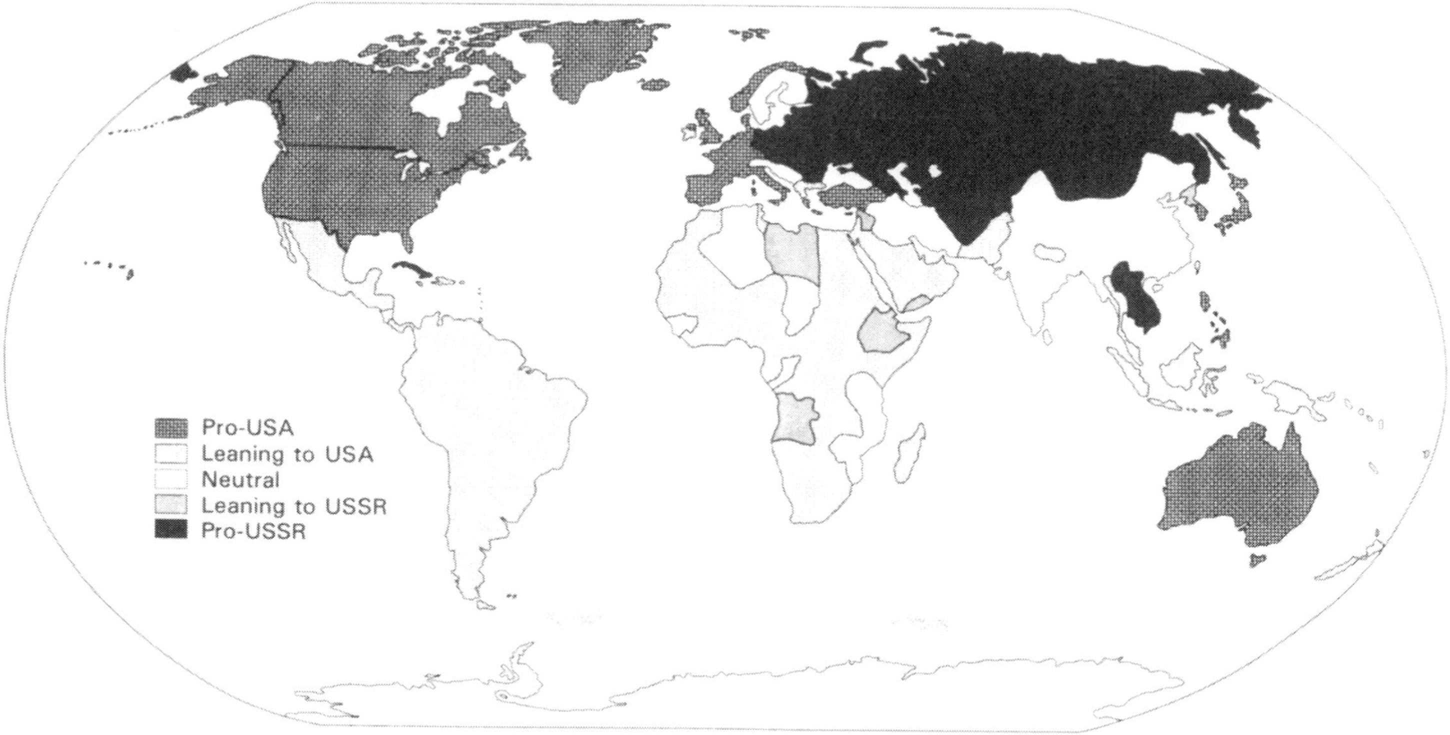

distance from the power cores, however. Obviously, competitive jockey-ing for position and leapfrogging play their part. Simple domino-like chains are absent. There is still evidence, however, that the force of pro-jected power does decrease with distance.

Even though the currency of maps of world politics is short-lived, the perception of world order in the form of a map of opposing camps does seem to inform the thoughts and words of the world's leaders. When they speak, the underlying image is of a globe of a few colours such as Figure 1.1. The vision of the world which provides a setting for some of the most important choices for the fate of mankind is probably a simple map, keeping the tally in the contest for influence by drawing lines of demarcation and containment, displaying the designs and fears of politicians.

To construct a realistic picture of the state of world affairs we need to draw objective maps of the force fields of the powers. This is necessary if we are to assess the reality of the image of duopolistic competition which lowers over thinking and writing and action on foreign affairs. It is essential for the testing of propositions about interlinkage between nations and the erosion of cold war fronts and curtains. For this we must seek consistent, accurate and unambiguous data on the components of power. With these we could draw a series of component maps and look for coincidence in their patterns. There is no proper way of combining these disparate qualities into an abstraction called power. The statistical manipulation of such data, seeking numerical principal components or common factors from a collection of numbers, is not an alchemy to discover the amalgam power, like a twentieth-century philosopher's stone. There is no such matter to analyse or reproduce. It is a verbal shortcut to refer to an array of weapons and means with which to act on others.

Interaction

The means of exertion between nations include armed might, diplomatic manoeuvres, control of trade and investment and the propagation of ideas. A complete account of the fields of influence in the world would involve measuring military, diplomatic, economic and cultural relationships bet-ween groups of people in different parts of the globe, both within, bet-ween and outside the authorities of nation states. Fairly complete and consistent data, however, are only available for nation states and the trans-actions between them. It is not possible to break down the places where choices and decisions resulting in flows are made finely enough to quantify the set of links between interest groups which are within a nation, on the one hand, or above and beyond it, on the other.

Consider economic transactions and data, for example. The buying and selling of goods and loaning of money are a matter of many deals between many different people in different places. The main purpose of officially recording these is to collect taxes levied on them. With no internal tariff barriers, there is no statutory call for data on flows between parts of a nation, only the individual agent's transactions are recorded and these are usually published only on a very aggregate basis to avoid revealing the business of a particular firm or person. The statistical record is published on a national basis, there being no administrative purpose for a finer geographic breakdown of economic data. Goods crossing a country's borders, however, destined for different foreign parts are treated differently. The collection of taxes does require the recording of flows to and from particular countries. Figures on international trade, then, provide a fairly complete picture of the movement of materials between nations, excepting those which are smuggled.

There is always a problem of attributing transactions or operations to an entity which makes decisions. Clearly, in matters economic we are dealing with aggregates of many people making choices resulting in flows between countries. Even so it is obviously silly to make statements about 'the nation' deciding to do one thing or another on this basis. It is the buyers or sellers of particular goods who are giving rise to these aggregations of transactions. It might be legitimate to talk of the wishes or desires of the nation on the basis of opinion polls, but these have not yet been elevated to a position of constitutional power anywhere. What is usually meant when the state is identified as having attitudes and making decisions is the *de facto* ruler, given the pressures which advisers, powerful people and politicians can bring to bear on the holder of this position. This is sometimes loosely referred to as 'the government'. In most nations, power is ultimately focused on one person, no matter what the constitution says. Thus, for the wielding of military might and diplomatic bargaining, at least, there is a person or small group with ultimate responsibility.

Force and diplomacy are in principle, then, a monopoly of government. These components of power are the singular responsibility of the political leadership of states. Since chains of command are not always well-defined and operable, however, and since the seat of power in the upper reaches of the leadership may be contested and shift even within a supposedly well-defined constitution, we cannot always be sure of the origin of these forms of power. There is, of course, some private enterprise in the use of armed force, including revolutionary and terrorist activities, private armies and clandestine, dirty tricks played by secret services for whom a wink is as good as a nod by way of authority. Aside from this, within the regular establishment of a state there may be several conflicting lines of military and diplomatic policy, by accident of competition, rather than cunning design intended to confuse the enemy. For example, there was evidently a serious dispute between the prime minister and foreign secretary of the UK on their respective sides of Downing Street, over the Falklands dispute. The struggle between the national security adviser, the secretary of state, secretary of defense, and UN ambassador for President Reagan's attention shifted the emphasis of US foreign policy and its tone considerably in 1983 and 1984.

In the 1960s there was a certain amount of academic activity measuring and analysing the transactions among nations. What these studies brought to light, whatever their initial intent, was the extent to which big powers intrude into the workings of other polities. These intrusions took the form of having colonial footholds close to other nations; intervening militarily; subverting the political order of other countries; making treaties; trading with and investing in other countries; manipulating the UN to condemn or encourage other nations; playing on cultural affinity; distributing new methods of doing things and broadcasting or insinuating propaganda.

Diffusion

If we start with the last two items on this list we are dealing with the spreading of ideas. This is sometimes done by broadcasting words and images for widespread consumption. Otherwise it is a more deliberate stream of influence aimed at some target population. With either widespread publication or more narrowly directed flows of sound, print, images, bribes, threats or personal influence, there is no necessary relationship between the quantity and quality of the message and the local level of reception. The geography of the impact of new things, ways of doing or ideas, depends not only on the intrinsic worth of the novelty, but also on the resistance which distance and terrain offer to diffusion and the local receptiveness to the message. For example, the elitist, more superficial form of Hinduism which was spreading through the Malay world around the time of Christ, gave way readily to Islam in the thirteenth century, while the ingrained spirit of Buddhism in Burma and

Thailand resisted its intrusion and the subsequent European influences. Although we might measure the acceptance of new implements or means of production or products in numbers, this does not necessarily register a change of heart or mind. Clearly, the propagation of ideas is not easily gauged and objective measurement of conversion to political ideals is not easily done in parts of the world where voting is infrequent, rigged or non-existent.

One factor which comes to bear on receptiveness to innovation or propaganda is cultural affinity between the sender and receiver of a message, idea or call to action. Long-standing cultural kinship between groups and nations, notably the use of a common language or possession of a common faith, might be expected to create and maintain friendly ties and favourable responses. Experience, however, calls this expectation into question. The first modern war, with an army of a million men involved, was a civil war between two sets of interests in the English-speaking, largely Protestant, population of the USA. In 1982 Hispanic sympathy with Argentina over the Falklands dispute with the UK fell far short of action. The supposed Soviet compatibility with Islam and its large Islamic population is not saving the USSR much grief in Afghanistan. Churchill hinted at the fragility of cultural similarity when he put it that the USA and UK were 'two nations divided by a common tongue'. Although the 'special relationship' which grew up after 1900 between the two was facilitated by their common heritage, this did not counter the British slide to third place after the French and Germans in importance to the USA. The style and doctrine of Margaret Thatcher and its appeal to Ronald Reagan may have altered the standings, however. Even so, ideological similarity is no guarantee of political warmth. Ideologies and religions seem capable of repeated fission into warring factions. Close proximity on the political spectrum does not seem to ensure amicable relations. The realm of ideas cannot be reduced readily to measurable and mappable simplicity in which closeness of characteristics implies friendliness or subservience. Since 1963 the USSR and China could hardly be placed in the same camp.

Diplomacy

The formal ties among nations of bloc membership, alliances and treaties are well known and mapped. These formalities do not, however, signify that the influence of the dominant partner is secure. Within NATO at present there is considerable doubt about the willingness of some members to follow the lead of the USA on several issues. Within the Warsaw Pact, although Poland's military rulers seek to placate the USSR, it is evident that a large part of the population is unhappy with Russia's influence, to say the least. The desire for departure from economic orthodoxy in the western satellites contrasts with the situation in the eastern satellites. Romania, for example, although it adheres to Politburo orthodoxy in economics, has been diplomatically nonconformist for some time. The lack of significance of the lower Danube makes this tolerable. All in all, if it can have little confidence in the Red Army for aggressive use, the leadership of the USSR must have even less trust in the military support of Poland, East Germany, Czechoslovakia, Hungary and Romania. In the Americas the Organization of American States (OAS) was once the vehicle for demonstrating the pre-eminence of the USA in the western hemisphere. It no longer functions in this fashion and shows signs of collapse under the strain of Central American and Caribbean tensions.

The margin of diplomatic dominance shifts back and forth in response to an erratic train of what in some cases are quite independent events. Trying to quantify diplomatic leanings with historical UN voting data may reflect the past disposition of military power and the fear of it; genuine ideological accord; dire economic necessity; greed or spinelessness. Whatever is the case, such behaviour will be subject to dramatic change with the fortunes of politics both domestic and international. Agreement or otherwise with UN resolutions is hardly a discriminating measure of national ideology. The US government does, however, intend to use a nation's UN voting record as the touchstone of loyalty. Jeane Kirkpatrick, then UN ambassador, persuaded Congress to attach a condition to the 1983 foreign aid package, requiring the president to examine each nation's UN voting record and to prohibit foreign aid to any country which did not behave itself in lining up with the USA's position. The particular target of this measure was Zimbabwe, which failed to condemn the USSR for shooting down a Korean airliner and sponsored a UN resolution concerning the US invasion of Grenada. It may be that in the future the leaders of those nations in great need may be persuaded to toe the party line more closely. The instability of their politics may, however, lead to radical switches in their loyalties.

The Military

Military presence is a far more concrete matter than diplomatic or cultural influence. The Pentagon and Kremlin publish maps of the locations of troops and weapons as evidence of each other's aggressiveness and their own purely defensive intentions. The International Institute of Strategic Studies in London produces an annual accounting of the geography of troops and armaments. Having men and arms in particular places does not always signify the same thing. Military presence is not necessarily a straightforward, positive measure of the influence of a powerful nation on the local population. Where force is placed may be a matter of local cause and the intended lessons may be for local consumption. On the other hand, the deployment may be a matter of global strategy with little intended relationship to conditions close by. Such a strategic placement of troops does hold the danger that foreign soldiers, or the danger of retaliation against them which would inadvertently kill local people, may upset an otherwise friendly populace. This has happened with US bases in Micronesia and is one of the dangers of any US deployment in the Gulf or Arabia. In some cases the location of troops is a matter of combined local and strategic considerations. In the numbers of Soviet divisions in Eastern European countries there is a correlation between the strategic need to deploy as far westwards as possible and the political need to counter local disenchantment with Communist Party rule by a display of Soviet strength. Strategic and political motives are combined in the absence of Soviet divisions in Poland, 4 in Hungary, 5 in Czechoslovakia and 20 in East Germany.

Putting troops or weapons in a certain place is often a political gesture with little military significance. The massive expenditures and intricate arrangements for nuclear weapons which have been laid out since 1962 and are intended in the future, are expensive political postures and bargaining chips. Quantitative or technical superiority would count for nought in a nuclear war where there can be no victory. In 1962 the US government demonstrated the military irrelevance of the Soviet presence in Cuba and in 1983 it did the same for the Cuban presence in Grenada. US marines in Lebanon or North Koreans in Zimbabwe count for little in any fighting, but they count for a lot in politics. Small contingents of soldiers provide a demonstration of support, show the flag and, in the last resort, provide a trip-wire guarantee of big power intervention in case of enemy action. Whether military presence in a region is a positive or negative index of the relations of nations is, however, open to interpretation. What is military assistance to the rulers of a country may be suppression for large numbers of the ruled. There are differences of opinion on whether Soviet soldiers in Afghanistan, or American soldiers in El Salvador, Guatemala and Honduras, represent an expression of friendship to those nations or an attempt to repress the legitimate aspirations of a significant section of their people. Given the games of postures and tokens being played, there is no necessary connection between the numbers of soldiers in places and their political significance. If anything, a large military presence often indicates an insecure relationship.

Economic Transactions

If we consider the economy as a weapon to be wielded by governments for political ends, then the power of this kind of intervention is also irregular and questionable in its impact. Time after time economic sanctions have not achieved their political goals. They failed against Mussolini in 1935, against Castro in 1959 and against Rhodesian whites from 1965 to 1979. These experiments in commercial leverage largely served to confirm the economists' faith in the substitutability of goods. We must wait to see the effect of the US embargo on Nicaragua.

Even though in the extremity of war governments have commandeered or co-opted the means of production and severed economic links, it is evident that complete control of production and consumption has evaded politicians. Wars come and go but property rights persist. In the Second World War cartel arrangements between DuPont and I.G. Farben restricted the production of material for US bomber nose windows, while patent arrangements between Bendix, British Zenith and Siemans constricted British production of aircraft carburettors. Even in command economies, control is incomplete and enormous shifts in national economies and international trade are largely independent of the state's power as such. Neither state ownership of the means of production nor stringent regulation of price, quantity, quality or location of output has been wholly successful in harnessing industrial might to serve the needs of the nation state's rulers. Even in the USSR managers have evaded directives to shift production away from the Moscow–Leningrad region in order to enjoy some advantages which do not impress central planners. The ways in which the interests of those who run and own the oil, car, steel, food, electronics, aircraft and construction industries impinge upon foreign policy are quite visible. The power of government to manipulate and direct the economy for power political ends is severely limited.

A more appropriate view of the international economy is not as a weapon of the state but as a largely independent network for satisfaction of material needs. Whatever doubt there may be about governments' economic powers and the efficacy of embargoes as a means of politicians getting their way, the links of trade and commerce are the fundamental components of any international community of interest. Even if anything less than wartime blockade have proved politically ineffective, nevertheless, on the positive side, flows of trade and investments are the most concrete expression of mutual interests. Although there are those who denigrate the benefits of some types of trade in terms of fostering undesirable dependency, there seems little doubt that a flow of goods balanced by a flow of payments does reveal the wants and needs of people expressed in purchasing power.

If we use flows of trade as a measure of the influence of countries on each other, we are changing the identity of the decision-makers in our analysis. Whereas we have considered the views and commands of one or a few political leaders as the trigger of events in other matters, in the case of the economy the power of choice is more fragmented and widespread. In the imaginary world of perfect markets, there would be no concentrations of buying or selling power and Adam Smith's 'invisible hand', or its latter-day mathematical derivatives, would guide all choices to a mutually beneficial outcome. In reality, economic power is more highly focused since advantages in costs of production and in facing the risk of an uncertain world accrue to large companies or investors. Individual businessmen can strongly influence the working of the economy and have amassed the clout to bend the actions of politicians to their will. This power is undermined by technical progress; by the fact that not all goods are produced efficiently only in large quantities; and by the ease with which people change their buying habits, using different products to meet their material needs. The spread of power in the economic realm is far more even than in the political one. In most societies, most adults have the opportunity to send signals on their desires and preferences for goods and services on a daily basis and get some reward for the use of their labour at least. Although their choices are constrained by limited incomes, limited information and limited access, they do get a chance to change their minds more frequently and to greater effect than in most politicial systems. Purchasing power is more potent, ready at hand and widespread in this world than voting power. Half of the countries of the world have no effective electoral system and in many others the polls are a sham. A case can be made for identifying the interests of the population of a nation with the flows of its economy, and thus tracing international influence through patterns of commerce.

The political leadership does sometimes express national interest in explicitly economic terms. President Carter openly identified US interests with Persian Gulf oil in 1980 and Reagan has reaffirmed this position. Without going to the deterministic extreme of dialectical materialism, it is reasonable to explain some part of the state of affairs between nations in terms of economic interaction. Although ties of culture, kith and kin may maintain uneconomically high level of trade, as Commonwealth preference did between Australia and New Zealand and the UK, the burden of distance may strain such bonds and encourage a more efficient realignment of exchange. Tariffs and trade barriers have been used not only to maintain old ties, but also to foster new ones, as in the cases of the EEC and its Eastern equivalent, COMECON. These too may distort production and trade from materially efficient quantities and directions. The role of conscious political choice in matters of trade and overseas investment denies us the prospect of coming up with a simple, mechanistic theory relating international politics to economic necessity. The arena within which the political choices can be made is bounded, however, by the limits of methods of production and the resistance which distance offers to exchange. What we have seen in the last century is an enormous expansion of the efficient scale of production for many goods and a diminution of the friction of distance. These changes have greatly extended the scope of political ambition. But they have not relaxed the restraints of scale entirely. The Soviet Union would be sorely pressed to subsidise another client with the problems of Castro's Cuba in Latin America.

From what has been written here it is clear that the current pattern of international exchange cannot be seen as resulting from a simple balance of economic or political forces. Nevertheless, it provides the least equivocal picture of shared interests among nations. The distortions of duplicity, fawning or show which colour public statements, UN votes and military postures, are less significant in commerce. Although flows of trade may be diverted by the interventions of government, to a great extent they are a recognition of mutual advantage between many people in the exporting and importing countries. Trade figures, then, will provide the basis for tracing fields of potential influence of one country on others.

The most fully and regularly recorded data are of visible trade. The International Monetary Fund publishes annual dollar values of flows of goods between countries. There is continuous coverage back to 1958. The other economic transactions between countries are not so well recorded or readily available. There are partial tables, like the US

Commerce Department's listing of US overseas investment, but no complete account of flows of services, funds and income. These are probably in fairly close harmony with the pattern of visible trade anyway. Trade flows, then, will provide the empirical observations from which to map the geography of influence and power.

Readings

For the measurement of transactions among nations see:

Brums, S. 'Transaction Flows in the International System', *American Political Science Review,* vol. 60, (1966), p. 889

Russet, B. *International Regions and the International System: A Study in Political Ecology* (Rand McNally, Chicago, 1967)

Savage, I., and K. Deutsch 'A Statistical Model of the Gross Analysis of Transaction Flows', *Econometrica,* vol.28 (1960), pp.551–772

The significance of UN voting patterns is considered in:

Rai, K. 'Policy and Voting in the UN General Assembly', *International Organization,* vol.26, no 3 (1972), pp.589–94

The global disposition of arms and men is given, mapped and analysed in:

The International Institute for Strategic Studies *The Military Balance 1983/84* (IISS, London, 1983)

Kidron, M., and D. Smith *The War Atlas* (Simon & Schuster, New York, 1983)

Soppelsa, J. *Geographie des armements* (Masson, Paris, 1980)

The use of trade as a weapon is treated in:

Renwick, R. *Economic Sanctions* (Croom Helm, London, 1982)

7 FLOWS AND LINKS

Trade Dominance

The mutual interests of people in different countries can be traced along the paths of trade. Potential fields of influence can, then, be mapped from data on commerce. Figures 7.1 and 7.2 show the countries whose import flows were dominated by the USA, the USSR or some other country in 1958 and 1982. The information of which these maps were based was taken from the International Monetary Fund's annual *Direction of Trade Statistics*. The earlier of the two, 1958, was the first year available after the Second World War and stood in lull before the disruption of Southeast Asia. The 1982 figures were the latest at hand. The tables of these volumes give, for each country, the value of goods exported to and imported from every other country in millions of dollars. Since 1976 the members of COMECON have not reported the trade between them and so, for our purposes here, the postion as at the last record is presumed to hold. Their exchanges with the rest of the world can be deduced from their trade partners' reports. To indicate the direction of dominance, it was presumed that influence comes along with imports. A nation is potentially under the domination of the country from which it imports the most. The reasoning behind the choice of the import flow is that political pressure is usually expressed or exerted as a refusal to sell goods to another nation. In the maps, countries are shaded according to whether the USA, the USSR or some other country was their largest source of imports in 1958 and again in 1982.

Both maps leave us in little doubt of the commanding global position enjoyed by the US economy in the 1950s and in the 1980s. Whether we consider territory or population, the superiority of the US is evident. In both years, distance does appear to be of significance in delineating spheres of influence, especially for the land power, the USSR. America's naval reach, as well as its productive superiority, give it a more far-flung domain. In both years the Soviet sphere is largely confined to Eurasia. In 1958 it was mostly contiguous to its own territory in Eastern Europe, Afghanistan, Mongolia, China and North Korea. By 1982 it had extended to Cuba. Its one outpost in Africa had shifted from Egypt in 1958 to Ethiopia in 1982. If the assumption of a stable pattern within COMECON since 1976 is true, then we would expect the Soviet economy

90

to have maintained its dominance in Eastern Europe. It has held Finland over the period and extended its reach to Yugoslavia in 1982. The major Soviet loss between 1958 and 1982 was of China. Afghanistan in 1982 was dependent on Iran for imports rather than the USSR, its dominant partner in 1958. There are obviously some other forms of Soviet influence involved here, although these are struggling for control.

In 1958 the afterglow of Western European, especially British and French, imperialism was still strong. This showed in the limited American and Russian influence in Africa, South Asia and Australasia, as well as on many Caribbean and Pacific islands.

The major change by 1982 was the reversal of China's trade connections. From Russian dominance in the 1950s, it shifted to neutrality in the early 1960s only to switch to dominance by the USA in the 1980s. Over the same period the USA won India from British adherence. The dominance of the USA in both of these cases is, however, marginal. Japan is a close second as a source of imports for India and China and its share of their trade is growing. The dwindling significance of the British Commonwealth connection has resulted not only in India, but also Australia and the Sudan, falling into the US sphere. In the 1960s and 1970s US influence in Southeast Asia waxed but since 1977 it has retreated, except in the Philippines.

Along the southern and western flanks of the Soviet sphere there has been a loss of US dominance in Pakistan, Iran, Turkey, Yugoslavia, Italy, West Germany and the UK. There has been a disengagement of the hegemonic spheres as measured by trade along an axis running from Liverpool to Lahore. This economic firebreak does have its political reflection in the emergence of a greater degree of political independence from the great powers along this principal corridor of history.

America and the USSR have traded places in northeast Africa with the USA relinquishing the Ethiopian highlands to the USSR in exchange for command of the Nile valley. The American dominance of Angolan trade in 1982 was a matter of Gulf Oil operating in Angola's oil-producing enclave within Zaire, Cabinda. According to *The Economist* (30 March 1985) Gulf is paying the Angolan government to have Cubans protect its installation against Jonas Savimbi's UNITA guerillas, who have the support of the US State Department.

In the western hemisphere, Cuba's trade attachment to the USSR is the most striking change, along with the emergence of a degree of independence from the USA in Brazil, Bolivia, Paraguay and Panama.

Figure 7.1: US and Soviet Trade Spheres, 1958

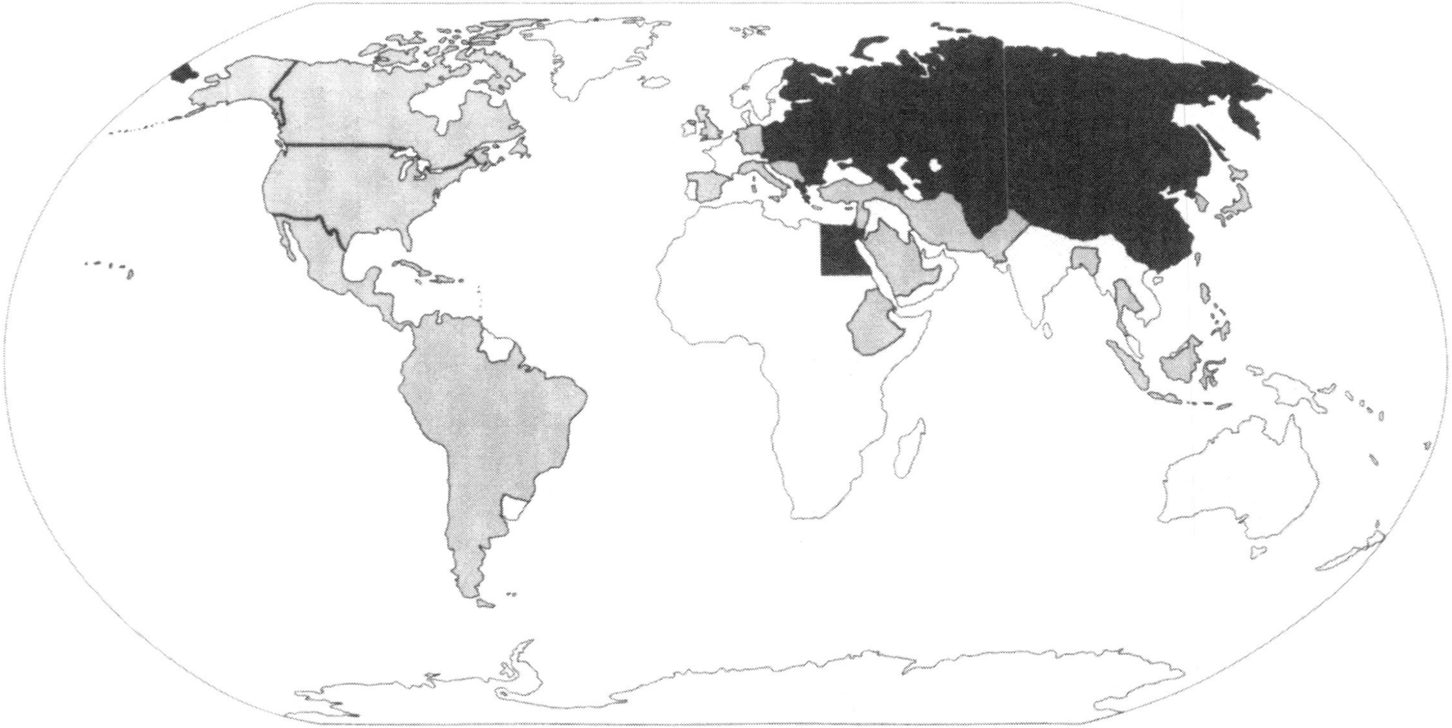

Figure 7.2: US and Soviet Trade Spheres, 1982

Links

This allocation of nations to one camp or the other or to the no-man's land between on the basis of where the greatest money payment for goods is made, presents a grossly oversimplified picture of power and political affiliations. In 1982 there were broad spaces where neither the USA nor the USSR prevailed directly. This included much of Western Europe, Africa, Southwest and Southeast Asia. This lumps a wide variety of circumstances into the same class of political condition. Some countries may be indirectly attached to one or other camp via an intermediary, while others may be truly isolated from either dominating influence. To give this intervening space more definition, the first order trade links between nations were established. Each country was connected to its major source of imports. Figures 7.3 and 7.4 map these trade-dominance links for 1958 and 1982 as lines connecting each country to its major source of imports.

By 1958 the USA had reached beyond its command of the western hemisphere. Across the Pacific it dominated Indonesia, the Philippines, Thailand, Taiwan, Korea and Japan. Across the Atlantic it reached into Africa (Ethiopia) and into Southwest Asia (Saudi Arabia, Israel, Iran and Pakistan). But the old imperial lines of the UK and France were still strong. The UK's trade dominance stretched to the Caribbean basin through West, East and Southern Africa and much of Southwest Asia. The imports of India, Ceylon, Malaya, Australia and New Zealand were still dominated by the UK. France too reached into the Caribbean, North, West and Central Africa, Indochina and Melanesia. Even the Netherlands and Portugal retained a few colonial links. Western Europe's trade was dominated by West Germany, with the UK standing out by dint of its reciprocal relationship with the USA. Beside all of this, the USSR's dominance of Eastern Europe, Egypt, Afghanistan, and its reciprocal relationship with China, looked somewhat puny. Already in 1958 the beginnings of Japan's expansion of influence could be detected in its dominance of Liberian imports. The northwest and northeast edges of the Atlantic basin, however, remained the main dominators of the world's trade.

By 1982, some new exporting centres of power had emerged. Japan dominated South and East Asia, parts of the Middle East, Panama and Liberia. Japan is, indeed, the major source of imports for the USSR after East Germany. In addition, Saudi Arabia, Kenya, South Africa, Australia and Brazil form minor nodes with two or more dependants. The reach of the UK to its former far-flung empire has diminished, although it maintains strong ties in West and Southern Africa. France has extended its dominance in North, West and Central Africa. France and Britain still have vestigial connections with the Caribbean and Pacific. West Germany continues to dominate Europe with some longer links to Tanzania, South Africa and Iran.

The Soviet Union has retracted its connection with China, and East Germany has been its greatest trading partner through the late 1960s and 1970s. An extension of influence to Yugoslavia and Cuba hardly offsets the diminution of its role elsewhere. Ethiopia is a poor replacement for Egypt. The USA has shifted its field of influence in the Middle East a little but, via Saudi Arabia, Egypt and the Sudan, has a major hand in affairs. The intrusion of the USSR into the Caribbean and the neutral main connections of Panama, Brazil, Paraguay and Bolivia, have reduced the US monopoly over the Americas. This was more than compensated for by the extension of domination to Australia and China.

Figure 7.4 does reveal a greater extent to US economic power than is evident in Figure 7.2. What is shown by drawing all of the links is the indirect lines of influence. We can see how the US dominates Saudi Arabia, Australia and Japan and, thus, indirectly has influence on their subsidiaries. Between 1958 and 1982 US command over the UK and its dependencies was lost. All of Europe, except for Spain, turned elsewhere for their main trading connection. There are, thus, some premier foci of economic affairs which are distinctly independent of the two great powers. These include West Germany, France and the UK. Although the UK has waned in imperial significance, it has at the same time become independent of the USA and more European. France has consolidated a territorially enormous economic foreland in Africa, and West Germany has extended its connections from Europe to Africa and Asia.

Thus, standing between the global reach of the USA and the more restricted trading empire of the USSR, we find strong and growing poles of economic power. The connections from Western Europe to Africa are more than mere colonial remnants sustained by the Lomé Convention. They display a geographic logic, linking the rural and extractive economies of Africa to the nearest major region of the urban, industrial world, in Europe. In the same fashion Japan connects with South and East Asia and the USA with the rest of the Americas. The connections of the USA to the Middle East are reflections of the strategic importance of its oil supplies and the material support of American foreign policy objectives. These political ties cut across the prevalent grain of global linkages and create an impression of economic intrusion much

Figure 7.3: Dominant Import Links, 1958

Figure 7.4: Dominant Import Links, 1982

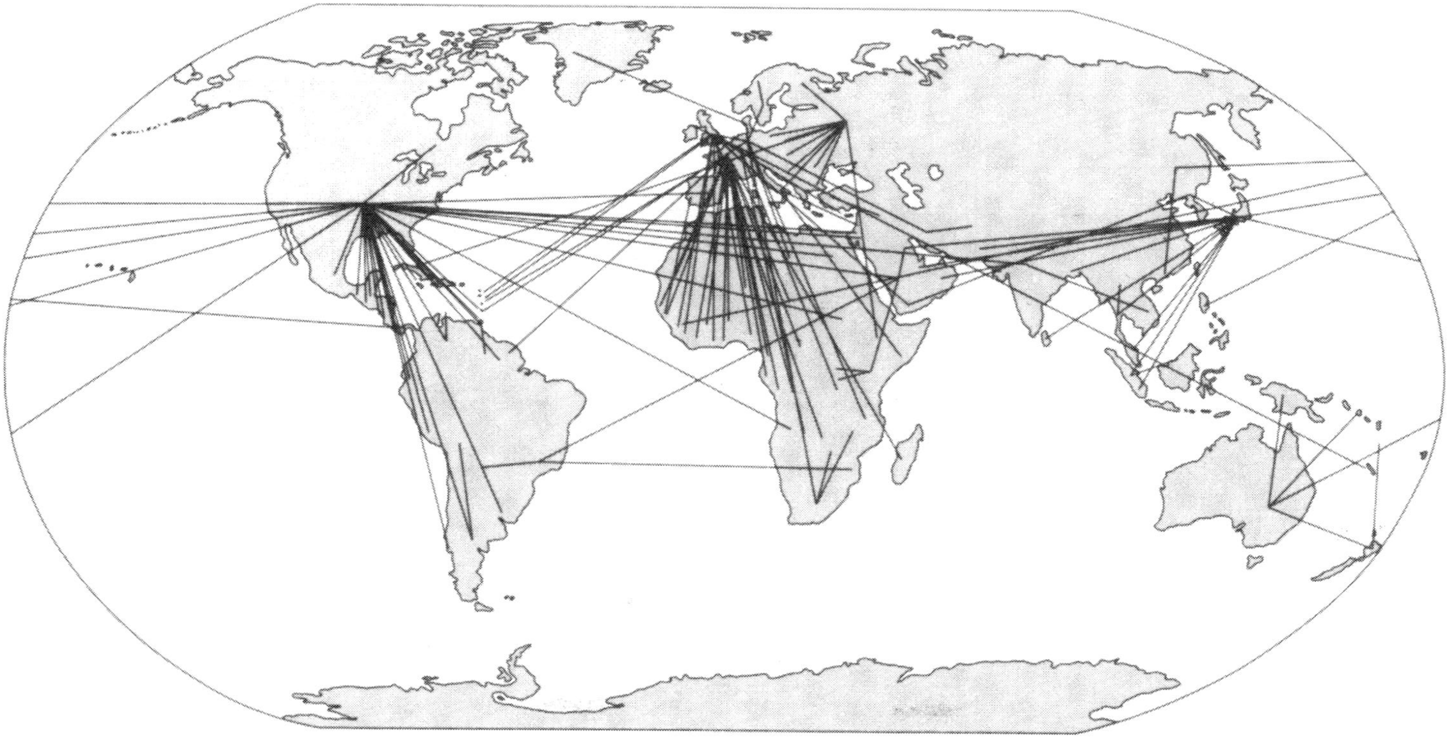

more pervasive than that of the USSR. For Europe, Japan or, indeed, the USSR to reach into Southwest Asian markets would seem more rational economic geography in simple, transport cost terms. The east-west, transoceanic links from the naval empires of Europe to their colonies have been partially replaced by the USA adopting a similar global role, while the more functional north-south links from the industrial north to the agricultural and mineral resources of the south continue to strengthen.

These superficial, first-cut examinations of world trade patterns as indicators of political relations suggest:

(a) the continuing global dominance of the USA;
(b) the limited extent of Soviet penetration of Eurasia, never mind the rest of the world;
(c) Japan's emergence as a focus of power;
(d) the integration of the Western European economy;
(e) the continuing strong links from Europe to Africa;
(f) a diminution of the US predominance in the Americas;
(g) the lack of articulated regional economies in Latin America and Africa;
(h) the emergence by 1982 of regional clusterings dominated by the USA, Western Europe, the Soviet Union and Japan, which reveal economic power to be more widely distributed than is power in the essentially bipolar world of nuclear weaponry.

The Economist (6 November 1982) pointed to the appearance of two major trading zones in the 1980s. Western Europe focuses a grouping which extends into Eastern Europe, Africa and Southwest Asia. The eastern portion of the US economy is engaged in this. The other, and more vibrant, zone draws the margins of the Pacific together with Japan as the primary focus. This includes the bustling economies of Taiwan, South Korea, Singapore and Hongkong. The western portion of the USA plays a major role in this grouping. Thus, the USA is a part of both of these trade zones but does not overwhelm them with its economic power. Within the USA, the conflict of interests between these two involvements has been a long-standing source of tension in US politics, and perhaps even within individual US statesmen. In general, however, the Pacific has long been a Republican ocean while the Atlantic was Democratic.

Interlinkage

The journalistic generalisation of *The Economist* was based on a consideration of the entire array of trade flows. The simple maps of linkage in Figures 7.3 and 7.4 define dominance so that a link may represent a small margin of superiority one way rather than another. The fact is that the links of trade between countries are many and widespread. If we were to construct a table with as many rows for countries as exporters as there are columns for countries as importers, the majority of the cells of this trade-flow matrix would have an entry in them. The International Monetary Fund statistics for 1980 show that for the 101 largest nations, with 10,100 possible flows between pairs of countries, there are 6,567 actual flows. Of all possible trade connections, 66 per cent are realised. The table of international trade is only one-third empty. This is a comparatively dense matrix of relations. Many of these are very small exchanges of less than a million dollars in value. To give these tiny, sporadic flows the same weight as the continuing, enormous exchanges, confuses the issues and hides the main outlines of the structure of international relations.

To filter out the less significant exchanges of goods for money, so as to reveal the chief sinews of economic interaction in the world, we need only consider those countries which imported a volume worth 1 billion dollars or more from a particular trade partner. Thus we limit ourselves to those countries which would have positive entries in a trade matrix of flows over a billion dollars in value. This reduces the number of countries to be looked at to 58. These include the anomalous cases of the oil-refinery island states of Bahrain, the Bahamas, Trinidad and Tobago and the Netherlands' Antilles. The matrix of flows of a billion dollars or more between these 58 countries is quite sparsely occupied. Only 10 per cent of the 3,306 possible flows of a billion dollars or more between the 58 countries occurred in 1982.

Looking for structure in this thin array of links reveals a weak hierarchical arrangement. This can be discerned by both a mapping of the data as in Figure 7.5 and by construction of a binary-flow matrix as in Figure 7.6. The map simply links countries between which there is a flow of a billion dollars' worth of goods or more. In the matrix a cell connecting two countries with a billion-dollar flow of imports to one from the other is filled in. Cells occupied by smaller or zero flows are left empty. As a starting-point the rows and columns of the matrix are ordered according to the number of import connections of each country, with the largest at the northwest corner. This produces the most compact array

Figure 7.5: Billion-dollar Flows, 1982

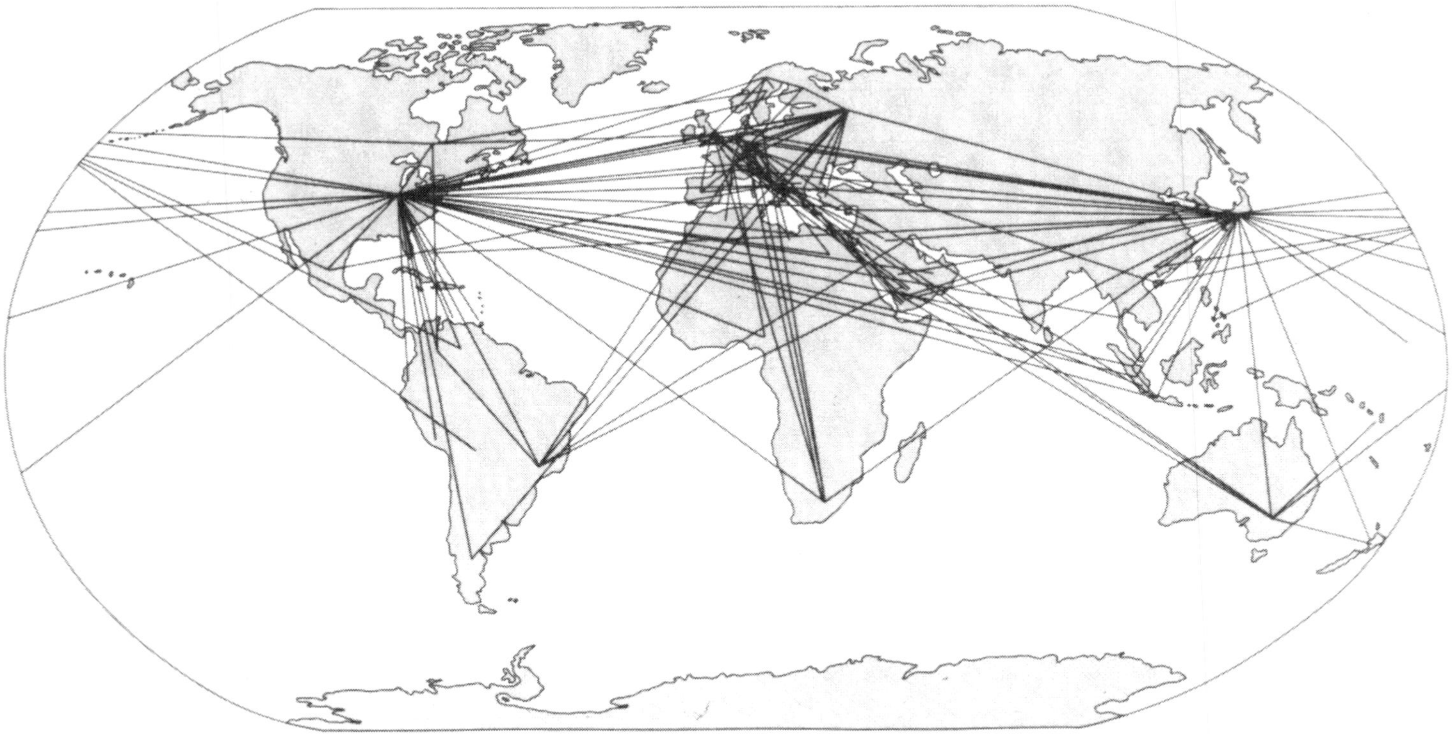

Figure 7.6: Billion-dollar Flows Matrix Ordered by Size

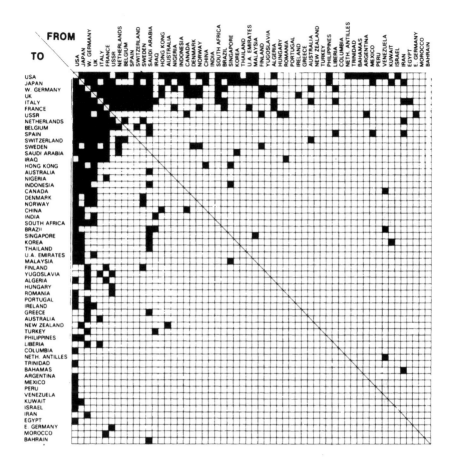

of full cells, with most flows confined to the northwest triangle of the matrix. Both this and the map point up a limited number of major foci. These are all weakly dominated by the USA, which has 34 input connections. Japan and West Germany follow strongly with 26 and 25 connections respectively. Then come the UK and Italy with 19 links, France with 17 and the USSR with 15. The Netherlands has 11, Spain and Belgium 10, Switzerland 9 and Sweden and Saudi Arabia with 8 apiece. The simplest numerical picture of this array is obtained by starting from the bottom and doubling up, adding up the number of countries with 1, 2–3, 4–7, 8–15, etc., import flows of a billion dollars or more. When we do this we get the listing given in Table 7.1.

Table 7.1

$b Flows	Countries
1	25
2–3	14
4–7	6
8–15	7
16–31	5
32 or over	1

This suggests several large foci connected with each other and with a set of mostly shared subordinates. The structure is much more complex than a simple hierarchy. Any hierarchical arrangement with each country connected to only one superior country will produce 57 flows between 58 places. With 329 cells occupied this is a much more articulate system. If we consider Western Europe, including the UK, as one, then this is the primary focus of trade, followed by the USA and Japan. The Soviet Union forms a slightly offset intermediate-sized focus, which is a tributary of Western Europe.

One way to cast light on the geographic structure of trade is to rearrange the rows and columns of the matrix of flows so as the have places in geographic order — that is, close to their neighbouring countries. Given that the construction of a matrix involves a reduction of spatial relations to one dimesion, obviously this requirement cannot be satisfied in full by having all propinquitous countries adjacent to each other in the list. Each country can only be adjacent to two others. For many clusters of more than two contiguous countries, contiguity on the map cannot be translated into adjacency of matrix rows and columns. The geographic requirement is, then, met somewhat loosely by seeking an ordering of the rows and columns which represents relative location by placing adjoining or nearby countries close together. 'Nearby' in some instances involves sharing a common ocean basin. Within the loose limitation of this geographic constraint, structure will be best revealed by trying to cluster flows as much as possible. This is done by finding an order of rows and columns which has the largest number of positive flows as close to the main diagonal of the matrix as possible. The diagonal represents the flow within a country and runs from the northwest to the southeast corner of the matrix. In seeking this clustering we come up against the problem revealed on the map. The world of international trade is essentially a cylindrical one, with trade flowing around the globe between the 60s of latitude. The Americas are nearby Western Europe and the Far East of Asia. This cannot be depicted satisfactorily on a two-dimensional display. However, after trying a number of orderings of

Figure 7.7: Billion-dollar Flows Matrix Geographically Ordered

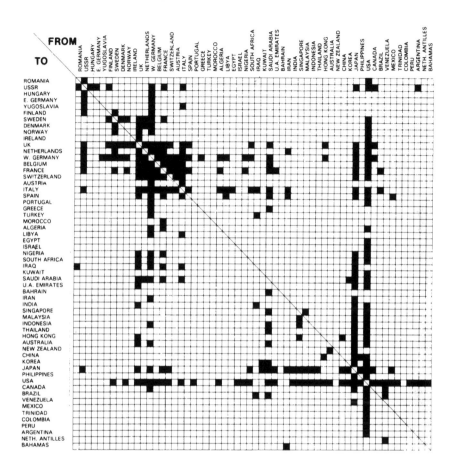

world regions, the one with the greatest clustering about the diagonal which retains some semblance of geographic order is that in Figure 7.7.

The overwhelming component of this pattern is the clump of intense interaction in Europe. This connects with the nearby minor focus in Eastern Europe and with a more far-flung foreland in Africa and Southwest Asia. Direct and indirect connections carry the linkage through this sparsely interconnected area, excepting the presence of Saudi Arabia, to the Far Eastern cluster centred on Japan. This then connects across the Pacific with the cluster of the Americas, which displays a strongly radial pattern focused on the USA. This in its turn links up across the Atlantic with Western Europe. The other striking feature of the figure is the global reach of the USA and Japan, as well as Europe. The strong interconnection in Western Europe is in sharp contrast to the essentially radial arrangements in the Americas, Eastern Europe and the Far East with few flows between tributary countries.

The limited and essentially subsidiary role of the USSR is much more striking in this picture of the world. Although the USA does individually have the greatest global reach, Western Europe collectively has at least as great a foreland and Japan is rapidly catching up. The overall pattern, a square of connections with a relatively empty centre, emphasises the fact that the main flows are between three major industrial regions with a string of flows to the undeveloped countries, which trade little between themselves.

To reduce the starkness of this picture and see if the structure deduced from the few major flows holds good when a broader spectrum of interaction is viewed, flows of over 100 million dollars were added to Figure 7.7 to produce Figure 7.8. This does generally reinforce the simpler picture, emphasising the overlapping nature of economic power fields, the absence of exclusive spheres of influence and bipolarity. The intensity of interaction between West European nations and their combined reach out to the rest of the world is manifest. The impression of the reach of Japan and the US is strengthened. The limited extent of the East European cluster, with Czechoslovakia and Poland missing, may be a product of inadequate data, since trade between COMECON nations is not available for 1982 and can only be inferred from 1976 figures. It is possible that the clump in the northwest corner of the matrix should be larger and fuller.

The poorly developed local trade of the southeast corner of the matrix, comprising the Americas, is a true reflection of conditions. What this broader display of the spectrum of flows does pick up is the emergence of the East Asian regional economy centred on Japan. This is strongly connected with Southwest Asia, where there is a distinct set of interacting countries less sharply focused on Saudi Arabia.

To reduce the resolution of this picture further by introducing flows of over 10 million dollars would reveal little new. It would simply fill in most of the matrix, leaving only the small negations of politics and geographical isolation showing. The economic linkages of the world are rich and widespread. If political influence is carried along with trade, then, it flows in many directions between many places. No country is totally isolated from others and a direct link to any country will lead through two or three indirect links to every other nation. The three great foci of international trade are in Western Europe, the USA and Japan

Figure 7.8: Hundred-million-dollar Flows Matrix

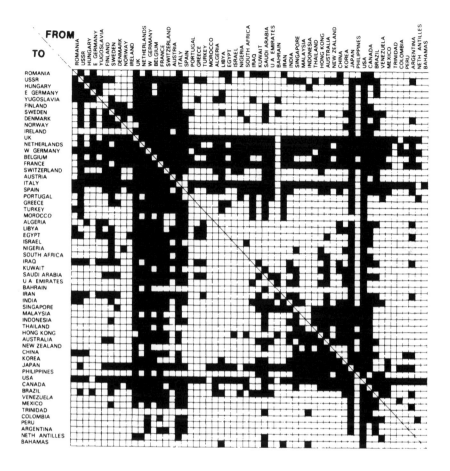

with the USSR as a sub-focus, tributary to Western Europe. The three foci interlock via their near-global coverage, with Europe's being the most massive. If we look at the direction of change in the pattern from our earlier deliberations, it is clear that the East Asian complex is the most dynamic element in the picture. If we extrapolate from the current position and the tendencies to change which we detected in the comparison of 1958 and 1982, it seems likely that by the end of the century, if we do not destroy ourselves or reach *dies irae* beforehand, a buffer of independent powers will have developed between the USA and the USSR reaching from the north cape of Norway through Europe to the Mediterranean basin, extending across the southern portions of Asia through China to Hokkaido's northern shore. The intensity of focus of this broad band of territory on the economies of Japan and Western Europe is already dissolving into a multipolar structures with a wider distribution of economic power, which, hopefully, will be accompanied by a wider spread of political power.

Readings

The classic work on the geography of trade is:

Thoman, R.S., and E.C. Conkling *Geography of International Trade* (Prentice Hall, Englewood Cliffs, NJ, 1967)

This is now dated and there is no good current description of the world pattern of trade. Newspapers, and especially *The Economist*, are the only decent sources of generalised and specific information. Data on trade are published as:

International Monetary Fund *Direction of Trade Statistics Yearbook* (IMF, New York, annual)

8 WAR AND PEACE

The Evolution of Nuclear Strategy

Since August 1945 military strategy and geopolitical reality have been torn asunder. The rift was complete by 1960 and when the USA and USSR faced each other with the power to deliver nuclear bombardments across the oceans. The explosion of nuclear bombs over Hiroshima and Nagasaki brought strategic bombing from the wings of the theatre of conflict to centre stage. Until that time bombs had been a dubiously effective weapon, especially when dropped on civilians. The leap of destructive powers unleashed by smashing the atom provided the rulers of the great powers with the capacity to eradicate mankind. These weapons could not be used to gain and exploit territory and people's allegiance, since they killed and destroyed beyond repair. The competition for superiority between power elites became a chillingly earnest exercise of technical ingenuity. Whether or not the existence of nuclear weapons has prevented a war between the USA and Soviet Union, it has ensured that if their competition comes to the last resort there is a danger of extinction for us all. There is no hope of either side winning anything worth while in a nuclear war. Nuclear strategists have lost sight of the purposes of rivalry. Victory is seen as the outcome of playing a game. The last one left standing wins. Notions from games devised as a preparation for real conflicts overtook reality and reduced the choice of life or death for us all to the level of a sport.

Questions of geography, of locations for launchers, of ranges, terrain, detection, response times and targets come into playing this game, but as far as our habitation of the earth is concerned the outcome of nuclear war would annihilate man and irrevocably blight the land. While the unholy threat of this fate hangs over us all, the leaders of the most powerful nations continue their competition for dominion in more discreet fashion. They intrude where there are local clashes of interest, to exploit the opportunity or forestall the opposition. This carries the danger that the buffers and proxies may be stripped away leaving full-frontal confrontation and the helter skelter path to death and destruction open.

The Austrians were the first to try to bomb a civilian population into submission, in Venice in 1849. The notion of long-range 'strategic' bombing of the enemy's homeland did not take full shape, however, till 1917.

The commander of the Royal Flying Corps in France, Trenchard, became convinced that the value of aircraft lay in their aggressive role. An Italian general, Douhet, provided a form for this in 1921 when he wrote a dismissal of tank and ground warfare, transferring conflict to the skies. Aircraft could be used to terrorise civilians, causing political collapse and cutting off military power at the source. Four years later, Liddell Hart advocated a combined tank lunge through enemy lines and air strike at their cities to paralyse their economy and polity. This provided Trenchard, who became the chief of the air staff, with a textbook for the development of a new form of war of attrition. The slog of the trenches was to be replaced by bombers wearing away the enemy's production. Through the 1920s and 1930s Trenchard demonstrated the effectiveness of aircraft in commanding territory in a series of operations in the Middle East. The popular fear of a direct air attack on British cities provided the political leverage to finance the growth of the RAF in this period. A rule of parity was established, ensuring that Britain could match the airforce of any country within striking distance of its cities. Trenchard saw to it that the expanding air fleet was geared for offence with bombers. Attack was the way to defend with aircraft. Goering's Luftwaffe in Germany was also built on offensive lines but integrated into the *Blitzkrieg* strategy in the fashion Liddell Hart had conceived of, to terrorise and disrupt the enemy in conjunction with a mobile, thrusting ground attack.

Billy Mitchell carried Trenchard's aggressive emphasis on long-range bombing back to the USA and inspired the instructors of the Air Corps Tactical School, who were later to become the airforce commanders of the Second World War. The doctrine taught at Maxwell Field, Alabama, was that the object of an airforce was to break the enemy's power and will to resist by hitting at the industrial capacity which fed military might. Attention shifted from military formations and defences to cities, factories, roads and railways. Mitchell's apostles extended the desired reach of US air power beyond the offshore bases he had advocated. They sought arrangements with likely allies from which to strike at potential enemies.

The political leaders of Britain and Nazi Germany were anxious to avoid air strikes against civilians, fearing the reaction of their own people to retaliatory attacks. The bombers were ready, however, and nearly a million civilians were killed and tens of millions made homeless by bombs. Despite the destruction, this was not a terribly effective weapon. The Blitz, as a period of history, is remembered with more affection than horror in England. The British policy of night-time saturation bombing of residential and industrial areas, which Churchill started in May 1940, was also militarily unsuccessful. It did not slow the German output of war material substantially and, if anything, provided a source of morale-stiffening propaganda for the Nazis. The one militarily useful operation was Tedder's plan to use the bombers to demolish French and Belgian railways and reduce German mobility in preparation for the Normandy invasion.

In April 1941 the fall of France started the US Air Force on a search for greater reach. The B36 was conceived as insurance against the fall of England and the consolidation of Germany's hold on the western rim of Eurasia. They were looking for a bomber that could deliver 10,000lbs of high explosives over a 5,000-mile radius. By mid-1943 it was evident that England would hold, but in the Pacific the success of the stepping-stone campaign to retrieve islands across the ocean was not certain. Thus, there was still a possible role for an intercontinental bomber. As the Japanese and Germans were driven back, the need for the B36 evaporated. Airforce planners were, however, already thinking about a war with Russia for control of the rimland. With doubts about the prospects for overseas bases and the extended reach needed to hit the USSR, the B36 looked a useful prospect, but it was still the case that strategic bombing was a marginal component of warfare.

The event which changed all of this was a military redundancy. The Japanese were probably willing to surrender by May 1945 and the Emperor was seeking Soviet intercession with the USA on Japan's behalf from June onwards. The decision to drop atomic bombs in August 1945 was based on US domestic politics and the desire to outstrip the Soviet Union in global influence. From a military viewpoint, the unleashing of nuclear explosive power gave the overwhelming advantage to the long-range offence directed at the enemy's heartland. In 1948 Omar Bradley voiced the threat as the doctrine of 'massive retaliation'. Any aggressive move of the Soviet Union in Europe wouuld be met by B36s and B50s carrying nuclear bombs to drop on cities. A small contingent of US troops in Europe would provide a trip-wire to trigger the nuclear response. Building the capacity to annihilate Soviet industrial and administrative centres would avoid the costs of attrition of conventional war. The US developed long-endurance aircraft, insitituted an airborne alert and cut back the use of overseas bases, using them mostly as refuelling stages. This was seen as the most adaptable arrangement with which to meet future contingencies, including Soviet possession of atomic weapons.

The USSR fulfilled this expectation with its first atomic explosion in 1949 and first hydrogen bomb in 1955. By the mid-1950s the USA and USSR faced each other with the power to deliver nuclear weapons across the oceans. The Red Army in Eastern Europe, which Stalin viewed as a deterrent to counterbalance the US nuclear threat, was now to be backed up by a fleet of Bisons and Bears carrying nuclear bombs. The USA, in fear of a 'bomber gap', responded by building B47s and B52s. The ground strategies of NATO and the Warsaw Pact converged to the common form of the forward active defence, with improved mobility and firepower, including tactical nuclear weapons, poised for speedy thrusts to carry the fight to the aggressor as fast as possible. The USSR incorporated medium-range missiles, with which to hit European targets, into its armoury. The Pentagon planners' view of Europe as the bat- tlefield for the forward defence of the USA in a fight which included the use of tactical nuclear weapons, did not please other NATO members, who had no control or even information on their deployment. Being defended to your own death is not an appealing prospect.

Soviet leaders and planners decided in the mid-1950s to concentrate on rockets rather than bombers for the delivery of their nuclear firepower. In August 1957 they fired their first intercontinental rocket and two months later demonstrated their global reach to the world by launching Sputnik. By 1960 the SS6 was ready for use. The rocket force was to be their main deterrent, roughly balancing American capabilities as they saw them, and allowing for a reduction of expenditure on ground forces.

The US military establishment panicked, fearful of a Soviet surprise attack through the 'missile gap'. Eisenhower tried to calm them by in- stalling 4,000-mile-range Jupiters in Turkey and Italy. He did not think that the Soviets could knock out Strategic Air Command. The influence of what Eisenhower called the 'military-industrial complex' was too much for him and the production of Atlas, Titan, Minuteman and Polaris missiles got under way. In 1961 it became evident to all that the Soviets had only a few intercontinental missiles in place and the gap vanished. But the momentum of the airforce and aerospace industry could not be diverted. US rocketry rapidly outstripped that of the USSR. The impend- ing superiority of American destructive power tempted Krushchev to try to outflank the US radar screen by putting missiles in Cuba in 1962. Failing in this, the Soviets turned to *détente* and building up their air defences.

Through the 1950s, as the Soviets built an arsenal theoretically capable of withstanding a US nuclear attack and striking back, massive retalia- tion became too dangerous a strategy. To hit the USSR with nuclear weapons to stop a conventional attack on Western Europe was to invite death for millions of Americans. McNamara introduced the doctrine of 'flexible response' to string out the confrontation, even when it involved nuclear weapons, so that there were several junctures for decision before the next step up in destructiveness was taken. Any Russian advance in Europe would be met first with conventional arms. As superiority of numbers told, the US would turn to the use of tactical nuclear weapons directed at troops. The steps up to theatre and intercontinental strikes would provide further cause for pause and diplomatic resolution of the conflict.

Through the 1960s faith in the planned precision assumed by the stance wore thin and it seemed more and more likely that rational response would vanish in a holocaust of destruction. It seemed to many people that there could be no victor in such a fight. By 1962 the USSR and USA had mustered enough explosive force to wipe each other out as recognisable societies. The view of nuclear weapons as the ultimate deterrent grew in the councils of the US govenment. It was increasingly assumed that the temptation to mount a pre-emptive strike would be restrained by the certainty of a nuclear counterattack. The Red Army's ambitions in Europe would be restrained by the horrible prospect of inevitable escalation from tactical to all-out nuclear war. In order to work, this US conception of deterrence by an implicit pact of 'mutual assured destruction' required that each side have the ability to strike back after a surprise attack and dash any hope of victory into the embers. For the balance of terror to be sustained, each side must be known to have an unassailable retaliatory force, removing the temptation to strike first. Any doubt about the vulnerability of this second-strike capacity carried the temptation to adopt a launch-on-warning policy and, thus, increased the danger of an accidental exchange. Whether or not the Soviet leaders ever felt themselves party to this mutual suicide pact, this strategy of menacing terrorism effectively rendered nuclear weapons of no military use. They only served to remove the temptation to play with nuclear fire from others.

Since they had always been trailing in firepower, the Soviets had taken some pains over defence, with improved radar, surface-to-air missiles and interceptors. Towards the end of the 1950s they had developed anti-ballistic missiles and there were a few sited around Moscow by 1965. Their ineffectiveness was, however, manifest by 1962 and the Soviet leadership began to show an interest in arms limitation. Nevertheless, the US military and aerospace industry had taken these evidently impotent defensive weapons as the grounds for developing multiple-warhead missiles. The Soviet defences were to be overwhelmed by numbers. By this stage the nuclear competition had departed from any relationship to the geography of political affairs. The contest raced onwards and upwards driven by the mad logic of technical ingenuity, systems analysis and schoolboy combative zeal.

While denying any intention of developing a first-strike weapon in 1971, the Nixon administration was deploying multiple-warhead rockets. These overturned the balance of terror entirely. Each missile could now destroy several enemy rockets at their launching sites. Single-warhead missiles could never achieve a killing rate of one or more enemy missiles per rocket. If they can be aimed well the multiple-warhead missile can knock out several enemy silos, giving the advantage to the aggressor. It is a first-strike weapon. After testing in 1968, the US had them deployed by 1970. The USSR responded with a test in 1973 and had such rockets in land launching sites by 1975 and in submarines by 1979. Some US strategy advisers saw this Soviet response as opening the defensive 'window', threatening US land-based missiles. The MX was proposed as a 'counterforce' to close this window and the idea of winnable nuclear war began to stalk the corridors of Washington.

The beginning of the notion of preparing to fight and win a war with nuclear weapons traces back to the founding of the RAND Corporation at Douglas Aircraft with General Hap Arnold's grant of 10 million dollars in 1945. This think-tank was the cradle of systems analysis. The old doctrine of massive retaliation was thrown into confusion by Soviet acquisition of intercontinental missiles. A 'countervalue' strike at the Russian heartland would invite the death of urban America. Obviously the Ruhr was not worth this. A new dispensation was preached by William Kaufmann of Rand. This was the doctrine of 'counterforce'. This has been the Soviet strategy from the outset of the contest. In American guise it meant that a Soviet thrust would be hit with tactical nuclear missiles. The Soviets might respond with longer-range weapons hurled at Strategic Air Command sites. They would not hit US cities for fear of a response in kind. In fact they have never aimed at these *per se*. Eventually the superior gunnery of the US would prevail and victory would only cost a few million dead on both sides, not hundreds of millions. Nuclear weapons would be directed solely at military targets, starting from the battlefield and working up to the silos. That the European battlefield would be blasted and burned beyond repair did not count in judging who was the victor. So the Rand creed held that nuclear war could be fought with some precision and won; that the main need was to have as impregnable a defence of the nuclear arsenal of the USA as possible and that the US needed counterforce weapons. Elements of this doctrine filtered into US nuclear policy through the 1960s, but it really became established after James Schlesinger was appointed Secretary of Defense. In 1974 he laid the groundwork for the counterforce doctrine by arguing that US defences were susceptible to a surgical excision of its silos by the Soviets, which would kill only 800,000 civilians. He proposed that any US president would then capitulate because a counterstroke would call down the end of American society. To fight a nuclear war the US needed to develop a counterforce weapon with which to bust hardened Soviet silos. The Kremlin scoffed at the idea that its weapons could be so surgically precise. More realistic estimates calculated a death toll of 20 million which, it can be argued, might lead a US president to an anguished doomsday response using his bomber and submarine forces. More recently predictions of dire environmental repercussions of even such a limited exchange and the lingering death of shattered societies damn the idea of victory in nuclear war.

None of this knowledge seemed to impress the exponents of preparing for a winnable nuclear war. By the end of Carter's administration the RAND counterforce doctrine was enshrined as official US policy in the strategic planning document 'Presidential Directive 59'. In June 1979 Carter gave the go-ahead to build the MX silo buster which Schlesinger had previously ordered designed. For Carter this was a political move to win over members of Congress who espoused the counterforce strategy in order to get Senate ratification of the Strategic Arms Limitation Treaty.

Counterforce called for building missiles to outweigh every element of the Soviet weaponry. The principal component of this hardware is a missile with ten warheads which can be guided to enemy missile silos with great enough precision to destroy them. When the MX is ready for action, the obvious interpretation of US strategy will be of a plan for a pre-emptive strike at the Soviet rocket force to deter conventional war. Although it is labelled a 'second-strike counter silo' weapon, the MX is patently something with which to hit first. There is no point in hitting empty silos. It makes sense only as a means of surprising the three-quarters of the Soviet nuclear armoury which is land based while it is in the ground. As a weapon for the counterstroke it would be destroying empty silos. Since it will present a first-strike threat to most of their weapons, the obvious Soviet response to the MX would be to hair-trigger them for a launch on warning, loosing everything to leave empty silos for the MX to smash. This would enormously increase the likelihood of finishing us all off by mistake.

The US government has not publicly disavowed previous pledges not to launch a pre-emptive strike, though it retains the threat of first use of tactical nuclear weapons. To pay lip service to this vow it was necessary to go through the motions of seeking a second-strike weapon deployment for the MX. This involved airing plans for a number of geographical shell games to ensure the survival of the 1,000 warheads necessary to strike back at the USSR's purported second-round missiles. Assuming that half of the MXs would survive a Soviet surprise attack in their launchers, this would require that 2,000 missiles with ten warheads apiece be deceptively deployed. It is by purest chance that the design total of 2,000 warheads would be enough to eradicate the 1,400 missiles the Soviets are entitled to under the arms limitation treaty, in a pre-emptive strike. After going through the grotesque jokiness of 'dense pack', 'big bird', 'hard tunnel' and orbital basing, it was finally determined that the MX, indecently renamed 'Peacemaker', would be located in 'superhardened' Minutemen silos in the Great Plains, whose locations are well known. Since superhardening counts for very little in the way of defence against accurate shots, this is obviously a first-strike deployment. The administration's public justification now is that the MX provides a bargaining chip in disarmament negotiations with the USSR. So far, however, Congress has authorised purchase of only 42 of these missiles.

Meanwhile, the administration has continued to speculate aloud in the counterforce vein, albeit in a more defensive tone. There was a burst of publicity along with the 1984 defence budget proposal in April 1983, for an array of 'Star Wars' killer laser and particle beam weapons under development. The Strategic Defense Initiative, as it is called, will involve spending 26 billion dollars over five years to get some ballistic missile defence gadgets in operation before the end of the present administration, making it very difficult for the next administration to opt out of the programme. The vision of an impenetrable defensive umbrella has proved somewhat leaky and a large body of expert opinion believes it could be only 95 per cent effective at most. Five per cent of the USSR's missiles could mess up American society badly. In addition, these defences would not be effective against cruise missiles and bombers. It has got to the point now where 'Star Wars' defences are being talked of not as a total shield, but as partial protection to screen some land-based missiles so as to defuse any Soviet hope of success with a counter-force, pre-emptive strike.

Europe

This dangerously unstable, escalating nuclear stand-off started as a series of bids in the auction for power over the northern plains and central uplands of Germany, leading to Europe's Atlantic shore. But the competition for mastery between the USSR and the USA was already of global proportions before the fall of Nazi Germany and Japan. The nuclear confrontation is the ultimate expression of this worldwide struggle and the particular significance of Europe has dwindled with the decades. The effects of any nuclear exchange would be oblivious of territorial niceties and would extend far beyond the European theatre. This has been recognised for a long time. In the last month of Eisenhower's administration the Pentagon's Single Integrated Operation Plan 62 was endorsed despite evidence presented that any US nuclear attack on the USSR would kill tens of millions of Chinese with fallout, even though they were not at war with the USA.

More recent calculations by Sagan and his colleagues predict that a fairly modest exchange of 5,000 megatons, about 40 per cent of the total deployment of the USA and USSR, would create enough dust and smoke to reduce temperatures in the northern hemisphere below freezing for months. This would destroy agriculture and kill most of the survivors by starvation and exposure, if they were not killed by the fallout. This nuclear winter could be triggered, they computed, by an exchange of as little as 100 megatons, less than a tenth of the firepower available. Whatever the damage to the plant, soil and animal life of the planet, it seems certain that this would destroy the tenuous economic and social links which sustain our imperfect ecumene. The extinction of the human species could not be ruled out as a possible outcome of even limited nuclear war.

Despite the fact that the conflict between the nuclear powers and the likely results of a nuclear war transcend the limits of Europe, it still provides the particular geographical setting for their nuclear strutting and threatening. The offically designated occasion for the use of nuclear weapons remains Europe. The USA made a commitment to shelter its NATO allies behind the power of its nuclear deterrent. The declared threat is that if the Warsaw Pact invades West Germany, the USA is willing to escalate the NATO defence from conventional forces to tactical nuclear weapons, to theatre weapons such as the Pershing, up to the ultimate use of intercontinental missiles aimed at Soviet missile silos. This policy of flexible response remains the avowed position.

This threat is less plausible if there is a gap in the sequence where the USA could opt out of the higher steps which would involve American cities. The perception by the advocates of the counterforce strategy that Soviet multiple-warhead missiles opened such a gap, led to the present reinforcement of US nuclear force in Europe. It was reasoned that these multiple-warhead missiles could be used by the Soviets in a pre-emptive strike to destroy USA-based missiles in conjunction with a conventional attack on Germany. If the home-based, missile element of the US nuclear triad were vulnerable to such an attack, then the threat of escalation rang hollow. Neither the allies nor the enemies of the USA could be expected to believe that any president would use the submarine and bomber legs of the triad in response to such a counterforce first strike. Missiles from submarines and bombers could only be used to attack cities in the USSR with any effect. This would invite a similar Soviet response, killing tens of millions of Americans. This would be too great a price to stop the Red Army marching into the Ruhr and the Low Countries.

Defence analysts' perception of an impending Soviet nuclear superiority in intercontinental missiles opening a window of vulnerability not only brought forth the MX, but also led to the decision to place Pershings and cruise missiles in Germany, Great Britain, Belgium, the Netherlands and Italy over the next couple of years. This is billed as a stop-gap to close the window and thus bridge the gap in the sequence of escalation until the MX is deployed in the USA and offers a credible counterforce threat. The Pershing II and cruise missiles are capable of reaching all of Eastern Europe and the western parts of Russia and are supposed to fill the breach between tactical and intercontinental weapons. By this means the US president assures his NATO allies that he will not take the opportunity of a firebreak to uncouple the USA from a European war and avoid an intercontinental exchange. There is not a wholly unfounded fear that Europe is seen as the battleground for the forward defence of the USA. If this involves nuclear weapons, Europe may be 'saved' to the point of destruction. Despite the placement of theatre weapons, it is still the case that a president of the USA has never explicitly said that he is willing to risk the death of metropolitan America for the defence of West Germany.

Looking for the geopolitical perception which fostered this unstable and potentially catastrophic competition, the fundamental blame would seem to rest with the fear of a communist scheme for world conquest, orchestrated by the denizens of the Kremlin. Western Europe is cast as the main item on this agenda. This myth, and the mission to conquer the 'evil empire' of communism which it inspires, does not stand up to close scrutiny as a realistic interpretation of Soviet foreign policy. Although it does seem that the Marxist-Leninist myth did provide an inclination towards change rather than the status quo in the foreign policies of Krushchev and Brezhnev, there is little evidence of it providing the driving force of Soviet actions in recent years, never mind inspiring a plan for world conquest.

The Russian empire shows signs of a slipping grip on its defensive

perimeter in Europe as well as in Asia. Feeling the strain of overexten-
sion, the main concern of Soviet leaders of late has been stability and
predictability in international affairs. Within the Warsaw Pact, although
General Jaruzelski seeks to placate the Kremlin, it is evident that a large,
organised part of the Polish population does not enjoy Russian constraints
on their range of choice. The Romanian government adheres to the Politburo line in economic matters, but has, for some time, gone its own
way in foreign affairs. The latitude allowed in different fields is largely
a matter of geographical position and, thus, strategic significance. The
governments of East Germany, Czechoslovakia and Hungary find it prudent to conform with Soviet political predilections, while achieving considerable economic independence of action. The historically founded
Russian fear of attack from the West could not brook too great a political
accommodation in that direction by its front line of defence.

Given grave doubts about the preparation and willingness of its own
army to engage in aggressive operations, the Soviet leadership can only
have less faith in the military support of East Germany, Czechoslovakia,
Hungary, Poland, and Romania for defence, never mind as partners in
conquest. The most obvious explanation of the Soviet presence, repres-
sion and machinations in Eastern Europe since 1945 is the USSR's defensive needs and fears. Fear of Germany in particular is seared into the
Russian collective memory. If this is so, then the nuclear confrontation
is militarily baseless as well as useless and puts all our lives at risk quite
pointlessly.

Whether the underlying process is perceived in the grandiose form
of Kissinger's geopolitical mechanics, the Nixon 'game plan' or an anti-
communist crusade, the prevailing American mental map is a
Mackinderesque projection divided between white and red camps with
a contested field of pinkish grey between. The image which sustains the
insanity of nuclear deterrence is of a violently aggressive Russian
heartland which must be held in check to preserve the social, economic
and political order of the world. To counterbalance this expansive source
of malignant energy, it is necessary to command the loyalty of those
who rule the rim of countries which surrounds this Eurasian empire.
Changes in government and local conflicts of interest are blown up to
global significance and other nations become mere dominoes in the
hegemonic struggle between the USA and USSR.

Although the most powerful leaders of Western Europe, including
the socialist Mitterand, endorsed the stepping up of US nuclear arms
in Europe in 1983, there has been a gradually increasing willingness to
go against the wishes of the US president and irritation with the arrogance

of US executive actions. European sensibilities have been bruised and fears aroused by being treated as nothing but a battlefield separating the two great powers. General de Gaulle's opening to the East in 1963, prophesying the establishment of a *détente* or even an *entente* from the Atlantic to the Urals was accompanied by open hostility to some American policies and finally by withdrawal from NATO in 1966. At the same time, West Germany's foreign minister, Willy Brandt, was pursuing Egon Bahr's *Ostpolitik*, seeking to allay Soviet fears and extend political and economic links eastwards. More recently, whether it be over gas pipelines from the USSR, trade or sport with the USSR or precipitate American action in the Middle East, Caribbean or Central America, the readiness to dissent by governments has increased and diffused. Spanish politicians expressed deep disquiet over American intervention in Nicaragua on the occasion of President Reagan's visit to Spain in May 1985. Even the British government, despite the Anglo-American special relationship and current ideological compatibility, protests aloud. Mrs Thatcher, whom Brezhnev called the 'Iron Lady', made her first opening gesture to the East in February 1984, with a visit to Budapest. Although it does not yet have a firm, abiding majority of the population behind it, the desire to be out from under the threat of the nuclear posturing of the USSR and USA is strong and growing. Sneering reference to Finland's position carries less and less conviction and its militarily neutral status looks more inviting to more people.

The trade data examined in the last chapter suggest that, economically, Europe is a separate focus of power with its own global reach, not just intervening ground between the duopolists. If politics and military affairs do eventually respect economic realities, then the nuclear standoff between the USA and USSR could in time be eclipsed by this independent entity that lies between them geographically on Eurasia's western peninsulas.

The Southern Flank

Despite rumours of their territorial ambitions among the Islamic lands to the south, the leaders of the Soviet Union are giving a good impression of imperial decline here. The futile brutality of the Soviet occupation of Afghanistan looks like the frantic grip of an empire in retreat, scrambling to hold on to a marginal province over which it has lost civil control. Local loyalties have fragmented the control network of the empire and defy ponderous efforts at restoration of the apparatus of the state. The USA has taken the opportunity to discomfit its rival by encouraging the mujahedin. The CIA have been buying Soviet-made arms which the Israelis captured from Syria and Egypt, in order to disguise American involvement. These have been delivered through Pakistan. The administration has expressed a desire to send aid more directly. The Chinese, Egyptian and Saudi Arabian governments also provide arms and finance to the insurgents.

What is ironic is that the American government is supporting groups of bellicose, religious fundamentalists whose political position is somewhat to the right of Ayatollah Khomeini. Their revolt against modernity in its Soviet guise is similar in vein to the religiously inspired casting off of the Shah and his version of progress, which had been underwritten by the USA. American wrath over rejection in Iran was visited on the Soviet government which found itself in a similar predicament, but in a region closer to its borders where its involvement was more direct, the potential loss of face greater and extrication more difficult.

The American government has, however, chosen to see the Red Army's 135,000 men stumbling among the ridges and ravines of the Hindu Kush as a threat to its vital interest in Persian Gulf oil. Counterguerilla operations in Afghanistan are postulated to be preparation for a march through Pakistan's Baluchi province to the shores of the Arabian Sea and, thence, to the Straits of Hormuz. Pentagon planners identify a second avenue of approach to Hormuz, and a commanding position over the oil traffic from the Gulf. This involves climbing the Elburz, crossing the Iranian Plateau and scrambling over the Zagros. These two routes are rugged, dry, hot, dusty and have inadequate roads. The local residents are belligerent and have traditionally shown an aptitude for discomfiting intruders. Nevertheless, if the Russians do indeed wish to control the oil resources of the Gulf directly, they will have to come at them overland. Direct conquest of the oilfields, rather than the sea lanes, would most likely involve an advance from Armenia through Kurdistan to Mosul and from there down the Tigris to the Shatt el Arab. The first part of this is over terrain as rugged as the first two lines of attack. The Kurds might be bribed with promises of consolidation and independence of Turkey, Iraq and Iran, but the Iraqi army is formidable, would be on its home ground and is versed, blooded and steeled for defensive action by its protracted war with the Iranians in the lower Tigris valley. This war provides an object lesson in the enormous cost, in men and material, of projecting military power over distance in this setting. The Russians are acutely aware of it and have assiduously avoided getting too closely involved.

The prospect of Soviet attack along any of these three avenues, if they could be called that, or Soviet manipulation of local politics and dissidence to secure a controlling interest in a quarter of the world's oil output, was the geostrategic premiss behind the US Rapid Deployment Force. Ths is an attempt by the USA to overcome the logistic disadvantages of extending its military power into a region where it has claimed a national interest but which happens to lie on the opposite side of the earth, 8,000 miles away. The grounds for this desire to increase the reach of its conventional firepower are not wholly convincing. Despite the CIA dissemination of misinformation, there is no good evidence that the USSR will be in such desperate straits for fuel so soon that it will be driven to conquer by thirst for oil. However, in 1980, reeling under the shock of the Shah's fall in Iran, Carter chose to reassert himself by marking out American turf in Southwest Asia. He declared that any other outsider trying to gain control here by direct or indirect means would be repelled, by US arms in the last resort. Reagan endorsed this and, more bluntly, threatened the USSR with confrontation if it moved into the Gulf. The extent of territory marked out by this imperative is not well defined. Although the focus was on the shores of the Persian Gulf and the oil fields, the geographical scope intended must be traced back to the 'arc of crisis' which Brzezinski dreamed up in 1978. This extended from the Horn of Africa to Chittagong. The interpretation is that the Carter declaration extended the promise of US support against violent internal or external opposition to the incumbent rulers of Egypt, Jordan, Saudi Arabia, Oman and Pakistan.

To keep this promise, nuclear deterrence will not do. No one believes that a US president would use tactical, never mind strategic, nuclear missiles to defend European and Japanese oil supplies from disruption. Nor is it likely that they would be used to turn back a Soviet invasion aimed at toppling the mullahs of Iran. The only credible threat was that the US would put men on the ground. The Rapid Deployment Force was conceived as a means of delivering sufficient soldiers and firepower to the scene of an invasion or insurrection to turn them off and supporting these troops for as long as necessary. Counting the potential cost after the commitment, this might place a breaking strain on the resources of the US military. The army chief of staff, General Wickham, has admitted that two geographically separated calls upon manpower and logistic backing — say a Soviet threat in Iraq and an outbreak between North and South Korea — would strain US forces to the limit. One prospect which Pentagon planners dwell on is of a Soviet invasion of Iran. To meet this the Rapid Deployment Force would take weeks to land the required 3½ army divisions (totalling 55,000 men, their weapons and vehicles), 1½ marine amphibious forces and 500 fighter planes. The continuing support of these men and machines with fuel, ammunition and food would surely strain US resources. The units employed have an alternative use as reserves for any European action and, so this theatre would be left thinly manned.

The foundation for the whole apparatus is the conjectured Soviet thirst for hyrdocarbons and, thus, obsession with having the oilfields of the Gulf. Close examination suggests that the more likely sources of violent trouble in the region are local rivalries. The real competitions for territory and power are between Arab and Jew; Maronite, Muslim and Druze; Shia and Sunni; monarchist and radical; the secular and the fundamentalist; nationalist and tribalist; and varieties of all these. The chief means of mustering energy to power ambition, however, even here where other identities battle for recognition, remains nationalism. Lesser and greater loyalties intrude as complicating factors. Although the Kremlin could stir up turmoil quite easily, the prospect of stepping in smoothly in its midst and taking over with hopes of an untroubled dominion is negligible. Extending its involvement with Islam into this hinge of the continents where Africa and Eurasia meet, has probably generated more grief than benefit for the USSR.

The motives for the USSR getting caught up in what the British army called the Middle East, were primarily to do with nuclear strategy, not energy policy and imperial land hunger. In the 1960s British and French withdrawal after the Suez crisis and Algerian War and a lack of American attention, left a power vacuum in North Africa and Southwest Asia. In the nuclear arms race the USSR was building up a blue-water navy to counter the US disposition of missiles in Polaris submarines. Denied a ready passage through the Dardenelles by NATO, invoking the Convention of Montreux of 1936, the Russians had to establish a squadron dedicated to service in the Mediterranean. Having decided against building aircraft carriers, it was necessary to find shore bases for the air cover for their ships. They set up first in Egypt. When thrown out of there they moved to Syria and then on to Libya. The premier objective remained the filling of this strategic need. Treaties were negotiated and soldiers and politicians were cultivated to try to secure these bases. If it did not work against the strategic aim, local communists were assisted in their ambition for power — as in Aden, for example. If it ran counter to meeting the strategic objective, or carried the risk of coming up against the USA directly, such missionary zeal was not indulged. This was the case with Syria. The Soviet abandonment of the PLO in Lebanon bears witness to this policy. The large Muslim population of the USSR and the accomodation between their clergy and the Marxist–Leninist state gave the Russians a reputation for compatibility with Islam. The Arab states were open to overtures. They were looking for a counterweight to the US commitment to Israel and for arms to fight Israel. The USSR came with a light tread among the sensibilities of local politics, bearing aid and arms.

Having spread its influence widely in the 1960s, the costs to the USSR of this extension of its sphere of influence began to mount in the 1970s. Familiarity breeds contempt. As Soviet numbers and demands grew, so local receptiveness waned. In Egypt, Russian advisers were sent back by Sadat in 1972 and a treaty with the USSR was torn up. Assad of Syria refused to sign such a treaty of friendship until 1980, continued to seek to regain Lebanon, thus risking the wrath of the USA, and continued to suppress home-grown communists. The diverse interests of its Arab protégés has put the USSR on both sides of several disputes from which it cannot hope to gain. Nevertheless, Kremlin perception of nuclear strategic needs led the USSR to push its presence further south.

To monitor and match the US missile-bearing submarines in the Indian Ocean the Russians sought bases for naval operations on the African coast. From a port on the Horn of Africa a fleet could not only patrol the Indian Ocean, but also the Red Sea and the Gulf. Barré of Somalia was doing battle over Ogaden with the Ethiopians, whose emperor, Haile Salassie, had the support of the USA. In return for military assistance to the Somalis the USSR got the use of the port of Berbera. Then, as if to illustrate how little control the USSR had in these affairs, in 1974 a radical, Mengistu, ousted the emperor and severed the American connection, turning instead to Cuba and the USSR for help against the Somalis and the Muslim insurgents in Eritrea. Barré turfed the Russians out of Berbera and offered it to the USA, looking for American and Egyptian help. This reversal of polarity runs counter to Soviet strategic interests and plunges it deeper into the political mess of Northeast Africa. It seems clear that in Ethiopia, as in Libya, local ambition is manipulating the great powers' involvement, rather than the other way around.

Qadaffi of Libya has spoken of his ambition, but it is for an Islamic state stretching across Africa south of the Sahara, not an extension of the Russian empire. His involvement in Chad, quarrels with the Organization of African Unity and with Morocco and Algeria over the Polisarios in the Western Sahara, along with his bankrolling of political violence by a wide spectrum of dissenters, are more of a liability than an asset to the USSR. A quirk of fate has put Qadaffi alongside the USA in supporting Morocco against the Polisarios, in his case to discomfit Algeria.

Soviet leaders have sought respectability in order to win a following among the quieter, predominantly rural nations of the world. There is an abiding desire for formal recognition of Soviet standing as a great power and to avoid a head-on collision with the USA. To these purposes the rulers of the USSR have sought to dilute conflict between Arabs and Israelis through the 1970s. In 1983 they stood well to one side as Israel invaded Lebanon and scattered the PLO. The Russians would like to be involved in any settlement between Arabs and Israelis, thereby legitimising their presence in the region. Such restraint does not endear them to their more ferocious clientele, including Syria, Iraq, Libya and the PLO. Yet it is the maintenance of these connections which would make the USSR an obvious broker for any real arrangement to hold war in check.

Africa South of the Sahara

In West, Central and East Africa people's lives are disrupted by drought, economic recession and the mismatching of their countries' boundaries, institutions and rulers' ambitions with the political and economic capacities of their societies. This great, raw, troubled, tropical region is removed from the mainstream of American and Soviet attention and strongly linked to Western Europe in cultural and economic terms. There are no great pickings here for superpowers, despite the opportunities presented by unsettled politics and mineral resources. The prospective costs of exerting influence and control far outweigh any expectation of benefit. There is, however, a bone for indirect contention between the USA and USSR in the subtropical south of Africa. Its commercial agriculture, mines, factories and towns and command of the seaways between the Indian and Atlantic Ocean, give it strategic and economic significance. The local circumstance which opens the door to outside intervention is the white minority's repressive maintenance of their supremacy. The hopes of blacks, raised by the improved livelihood available to some of them, are stubbed against the social and political barriers of apartheid. The competition for power and resources between the dominant white tribe, on the one hand, and coloured, Asian and a variety of black interests, on the other, provides distinct sides for the hegemonists to take. The Reagan administration has opted for military rather than moral superiority and has shown increasing favour to the white, nationalist holders of power. This has provided military victory over communist-backed forces, maintained the status quo, provided friendly control of the Cape and access to strategic mineral resources. Although lip service is paid to moving the Afrikaners eventually to more acceptable racial attitudes and behaviour, it is difficult not to see an unspoken racial preference in this choice of sides.

The US government has eased restraints on the sale of military and police equipment and computers to South Africa. The resulting trade included a shipment of cattle prods, a high-tech version of the sjambok. American companies have been allowed to provide services to the South African nuclear power authority. The US Department of Commerce has opened up shop in Johannesburg to promote trade. All of these actions have been taken to strengthen a bastion against the red tide of communism. In particular, Cuban soldiers in Angola are deemed to be a threatening extension of the evil empire. The US government condoned, or, at least, did not condemn South African incursions into Angola and assistance to Jonas Savimbi's UNITA guerrillas. These have pushed their operations within striking distance of the capital, Luanda. This severely frightened the government and strained its Cuban allies. In another of Africa's political ironies, Cuban soldiers are being paid by the American Gulf Oil corporation to defend its oil installations in Angola against UNITA attacks, which are being encouraged by the US State Department.

Lack of censure and the positive gestures of the US government have been taken by South Africa as a licence to strike into the territory of its neighbours to wipe out South African blacks in exile, organised as the African National Congress (ANC), who seek to overturn the nationalist regime and apartheid, or members of the South West African People's Organization (SWAPO) who seek to sever South Africa's hold over Namibia. The unfounded fear of a concerted black crusade against white South Africa by the people of the frontline states is allayed by a policy of disrupting their affairs. In 1981 the South African army hit SWAPO targets in Angola. In 1982 a campaign of sabotage was directed against ANC establishments in Zimbabwe. Raids were carried out against exiled dissidents in Lesotho in late 1982 and early 1983. The ANC offices in the Mozambican capital, Maputo, were bombed by the South African airforce in 1983. Prime Minister Botha has likened the mission of the Republic in Southern Africa to that announced for the USA by the Monroe Doctrine in the Americas. 'Stability' is to be promoted and 'democratic' forces are to be strengthened against communist subversion by a policy of destabilisation of their neighbours.

In the short run, South Africa has gained the upper hand and its neighbours cower before its military and economic might. This has cost the South African government 10 per cent of its annual budget and a considerable loss of life. Having reduced the Angolan government's territory to the northern part of the country around Luanda and having made them utterly dependent on their Cuban trip-wire, in March 1984 South Africa agreed to a cease-fire and disengagement of its forces in Angola. Namibia is occupied by 21,000 South African troops with 9,000 black Namibian allies, arrayed against 6,000 SWAPO fighters in Namibia and Angola. In February 1983 General Malan, the defence minister, threatened to step up South African support of anti-government forces in frontline states unless they signed non-aggression pacts with South Africa. Lesotho and Botswana are entirely at South Africa's mercy. Kaunda of Zambia has gone so far to talk of South Africa joining the Organization of African Unity. The lever of support for the 10,000-strong Mozambique National Resistance movement has brought Machel of Mozambique to heel, signing a non-aggression pact with South Africa in March 1984. Only Mugabe of Zimbabwe has not submitted. Zimbabwe, with its significant white minority, has been the main recipient of American attention among black states in Southern Africa. Relations between Mugabe and the USA have frayed badly over the last year or so. Mugabe refuses to negotiate with South Africa to betray the ANC and sees US actions as effective permission to South Africa to pursue the exiles of the ANC across its borders. Zimbabwe's refusal to condemn the USSR in the UN over the shooting down of a Korean airliner and reproof of the US invasion of Grenada brought forth official expressions of chagrin from the State Department.

The primary object of South Africa's violent adventures abroad is not any foreign threat to the security of white society but the opposition that arose within the country as a result of the brutal injustice of the apparatus of apartheid. Given the efficiency of the South African police and spy system, this could organise only in exile. It must seem to the black population of Southern Africa, at least, that the USA is encouraging South African military aggression aimed at protecting white supremacy. To the Reagan administration, South Africa's successful intimidation of its neighbours was a victory in the global contest with communism. To Pretoria, communism and opposition to apartheid are one and the same thing. To the Afrikaners, the prospect of a communist 'total onslaught' implies a black race war threatening to swamp the whites.

The short-term win against 'communism' which can be marked up as the South African military reduce their black neighbours and residents to docility, may have longer-run repercussions which are not so comforting to the USA. It does provide a constituency for any Soviet ambition to expand Russian influence in the region. Denied the military upper hand, the USSR can seize the more virtuous role of encouraging the aspirations of the people of the frontline states and of the 83 per cent majority of South Africa's population which is non-white. Blacks, 72 per cent of the population, are either confined to the scrubby dryness of the 13 per cent of the country set aside as tribal homelands, or are used and constrained as what amounts to a latter-day helot class in the white, urban, industrial economy. The Afrikaner establishment exploits tribal differences and the divergent interests of town and country blacks in order to divide and rule. The coloured and Asian members of the population (10 per cent) are being offered the bribe of constitutional recognition in harmless representative chambers. Urban blacks are fed promises of homeland status and a degree of autonomy within their city neighbourhoods. The feudal crudity of tribal leaders is furnished with money, arms and military assistance to preserve its divisive influence. No matter how heartfelt the hope of American statesmen that their close relationship with South Africa will enable them to guide society there through a peaceful evolution to multiracialism, to much of the rest of the world benevolence to the Afrikaner government is support for white supremacy.

The Americas

If the American desire to score immediate military wins in what they view as a worldwide match against the Soviets and their Cuban henchmen wells up in Africa, it is raised to fever pitch in their own backyard. Here, however, the Cubans are close to home also. In addition, the forested mountains and intensely occupied agricultural patches of Central America are a much better proposition for insurgent guerillas than the dry, lightly peopled plains of Namibia and Zimbabwe. Not only is the terrain conducive to insurrection, but the political and economic conditions of the region are atrocious enough to drive people to seek redress with violence. The mainspring of this turmoil is the oligarchic stranglehold over resources which persists in many parts as a hangover of Spanish colonialism. Since the 1890s the USA has usually supported these oligarchies in its interventions in the region, as a defence of the status quo and of American investments in the region. This inclination has been reinforced by the unreasonable fear and hatred of revolutionary movements that warps the geography of international politics into a battle for the world with the USSR.

What should quiet this fear is ample evidence of Cuban independence of action. The spreading of Cuban influence with a policy of '*internacionalismo*' is not a simple extension of the Soviet domain in an hispanic guise. Although they are heavily dependent on the 4 billion dollars they get from the USSR every year, the Cubans do take their own line of conduct. In Grenada, for example, Castro had been backing Maurice Bishop while protégés of Moscow were responsible for his overthrow and death. Castro has been advising poor nations to look to the USA and Western Europe for aid, rather than to the USSR. He has been prompting moderation among the ruling Sandinista council in Nicaragua and among the guerrillas in El Salvador. Since their big push in early 1981, the flow of assistance and arms to the insurgents of El Salvador from Cuba and Nicaragua has been minimal. The failure of the best efforts of the considerable American spying apparatus to come up with proof of any substantial traffic, bears witness to this fact. The military powers of the guerrillas and popular support for the movement are evidently strong enough for resistance to be largely self-sustaining.

Cuban prudence was increased by the US invasion of Grenada. Their continuing fear of a US attack on Cuba has deepened. The Cubans have expressed misgivings that guerrilla success in El Salvador will prompt further US military action, even to the extent of invading Cuba. The Cuban response is to pull back from its military involvement. Castro has told the Nicaraguan government that it cannot depend on Cuban help in the event of direct American military action. Nevertheless, the Cuban government intends to continue to send teachers and technicians abroad in a peaceful diffusion of its influence in the region.

By contrast with this more prudent Cuban strategy, the Reagan administration has upped the ante in Central America and the Caribbean. In September 1983 an under-secretary of defense proclaimed the goal of US policy to be 'victory for the forces of democracy', even to the point of overthrowing the Sandinista government of Nicaragua. Every military move by revolutionaries is to be met with a counterpunch in a policy of 'symmetry' of violence, treating the region as a whole. The success of guerrillas in El Salvador was to be met by the increased subsidy of Contras in Nicaragua. In April 1985 the Secretary of State banned trade with Nicaragua in an admittedly empty gesture of hostility. The nastiness of their protégés or the ills of societies are given minimal weight in taking sides. The ultimate object is to keep the map of the Americas clear of communism and what are portrayed as Soviet tentacles feeling at the soft underbelly of the USA.

South America's variegated human geography does share a common economic fate. All of these countries are deeply in debt to international lenders. The world recession has reduced the export earnings that they need to service their debts. This, along with austerity measures imposed to meet debt payments, has increased unemployment and decreased real incomes. As a result, imports from the industrial countries of the world, especially the USA, have been sharply curtailed. As this contributes to US unemployment levels, the pressure builds up for protectionist measures directed against Latin American manufactures. Although there is a wide disparity in the general level of material well-being, all of the nations of Latin America are faced with large numbers unemployed and reduced purchasing power. This comes close on the heels of the expectations aroused by the economic surges of the 1960s and 1970s. Here we have a prescription for political turmoil in a continent where the usual mode of government is military dictatorship. The claim to authority of such a regime is that it provides efficient government without the social disruptiveness of politics and provides for increased well-being. Inability to deliver calls the legitimacy of this style of leadership into question. The tone for the decade was set in 1982 by the Argentinian junta's failure in a traditional recourse of self-appointed regimes faced with domestic trouble — military adventure. Its ineptitude over the Falklands wiped out any claim to authority on the base of competence. These warriors could not even carry out their own business of war effectively, never mind managing a society.

In their scramble to maintain their footing, military rulers have sought friends and purchase where they might — and not just with the traditional source of patronage, the USA. The international links and politics of Latin America have become more widespread and complicated. Argentina's major export customer is now the USSR. The Argentinians made up the deficit in Soviet grain requirements resulting from the Carter ban on sales following the Soviet invasion of Afghanistan. This trade relationship generates some political exchange and there are indications that the USSR helped out with intelligence at least in the Argentinian battle with Britain over the Falklands. This, even though one of Galtieri's justifications for possessing the islands was in terms of the need to keep an eye on Soviet submarines in the South Atlantic. He embarked on the quest for the Falklands/Malvinas in hope of a lenient attitude and, indeed, help from the USA. The American government had sought his friendship and enlisted his military involvement in plans directed to overthrow the Sandinista government in Nicaragua.

The USA is no longer the sole source of weapons and the hardware of war for aspirants to expanded power in South America. Peru has purchased arms from the USSR and Brazil has become an armaments producer in its own right. What used to be an American monopoly and powerful means of persuasion is broken. The USA is having to compete actively in the influence stakes.

The Reagan administration has not, thus far, taken the lead over the crucial issue for all of Latin America — debt repayment. The commanding position of the US government might enable it to ameliorate the economic and political turbulence generated by these debts and win lasting allegiance. But the administration has not used its leverage to establish generous terms for repayment, nor has it extended its aid to offset the cutbacks which are eroding the social fabric in Latin nations. US budget deficits exacerbate the problem by sustaining high interest rates, inflating the service payments on the debts. The administration's inclination, faced with domestic unemployment, is to protect US steel, footwear, leather goods and cement producers from Latin American competition. This bodes ill for hopes of recovery which must lie with increasing international trade. Paul Volcker, the chairman of the Federal Reserve Bank, and the private banks have, however, held off disaster with a series of *ad hoc* arrangements.

The president and his closest advisers are preoccupied with Central America as the theatre closest to home in which their global struggle with communism and Soviet imperialism is playing. They show little interest in grasping the opportunity to engineer a stable economic and, thus, political order in Latin America as a whole. The signs of unrest and potential violence from the right and left are manifest. The Sendero Luminosa guerillas in Peru are matched by rightist rumblings in Brazil and echoes of Peronism in Argentina. The frequency of riots, looting and bank robberies is picking up.

The Far East

The nuclear strategms of the USA and USSR and the global pattern of industrial growth and change have conspired to bring the eastern rimland of Eurasia back into the limelight in the 1980s. Whereas in 1970 trans-Atlantic trade was double that across the Pacific, now the volume of trade between Asia and North America nearly matches that flowing to and from Europe, Africa and the Middle East.

Meanwhile politically, a cold front has intensified along a line from Singapore to the Bering Straits. The enormous presence of China signifies very little in this conflict. It is a matter of Japan, South Korea and the USA lining up against an increased Soviet deployment of force on its eastern shores, directed at the USA not China.

The decision taken in 1979 to place Pershing IIs and cruise missiles in Europe threatened to reduce the warning the Russians would have of nuclear attack. These weapons could strike at the heart of Russia only six minutes after being fired from Western Europe. This was a frightening prospect. To counter the threat, the Soviets turned to the Pacific. They are building bases for submarines around the Sea of Okhotsk. Submarines based here can shelter in the Pacific deeps and threaten the western USA with a similar quick strike with missiles. This might head off the temptation for the US to attempt a fast strike from its European launch sites. The Soviets also stationed Backfire bombers on the south Siberian coast and built a military base on the Kuriles only miles from Hokkaido.

In September 1982 the Japanese government agreed to the basing of US F-16s in northern Honshu. The current Japanese prime minister, Yasu Nakasone, has a long record of pressing for a build-up of Japanese military might, to assist the USA. Coming to power in October 1982, he pledged Japan to control the straits that approach the Soviet Pacific naval bases. In January 1983 Nakasone made the first official visit of a Japanese prime minister to Seoul, South Korea. Even though his more militant attitude has not overcome the deep popular antipathy to preparation for war, it has won the admiration of President Reagan. Yasu is Ron's favourite foreigner. Although Maggie Thatcher is ideologically sound, she has displayed a sharp independence in foreign affairs. Nakasone and Reagan have even contrived some accommodations in the economic sphere. Japan's superiority in production technology and long-term economic management has left its competitors floundering. Recession and unemployment in key industries and a big trade deficit with Japan have built up the pressure for protection. Some of this has been headed off with Japanese market-opening measures and *ad hoc*, mutual curbs on competition, protecting the US car industry and Japanese citrus and beef producers, but Japan's growing trade surplus with the USA remains a source of contention.

By the start of the 1980s, China's rulers felt less threatened from abroad. The USA had sought an accommodation with them. It was evident that the military build-up in the east of the USSR was pointed at the USA not China. In general they have drawn back from confrontations and entanglements in the vicinity of their own borders, never mind further abroad. Chinese political energies are taken up with internal stresses and the struggle between continuity and change in Chinese society. There are strains of contradiction set up between the rising material aspirations of farmers, artisans, labourers, technicians and managers, on the one hand, and a system of government designed to promote collectivism, on the other. The communist establishment is patently outmoded and is showing the cracks of inflexibility in the face of a changing social setting. Deng's attempts to dismantle the Maoist edifice and replace it with something more robust and lasting are grinding to a halt, caught between the intransigence of the old guard and the impatience of the mass of individual ambitions. Openings for exchange with the USA, Japan and Europe, has fired desires which are at odds with the rhetoric which sustains political power. The need to resolve these internal contradicitions takes up the attention of China's leaders and foreign policy is subordinate to this goal. The missionary zeal to lead the world through revolution that flared through the 1950s and 1960s has dwindled to the merest flicker and during the last decade China has drawn back its direct military support of revolution abroad.

The dealings of China and the USSR are polite and restrained. There is some evidence that the Russians plan to withdraw troops from Mongolia. The Mongolians were most displeased with this development and brutally expelled many ethnic Chinese to China to register their dissatisfaction. China accepted this with equanimity. The foreign ministers of the two powers continued to meet and talk.

In the cases of Hongkong and Taiwan, the Chinese government and the current rulers of these territories have difficult tasks of reconciliation. But if the current of craving for commercial freedom and affluence becomes the mainstream of Chinese politics, the gap between the mainland and its appendages will close, in terms of expectations at least.

Although the USSR is an ally of Vietnam and is thus entangled in Kampuchea, there is little prospect of this bringing it lasting imperial advantage in Southeast Asia. The US government would find it difficult to muster support to venture back into Indochinese affairs and so there is little danger of great power confrontation here.

The autocratic military machine of North Korea continues to desire the South, with its industrial success and affluence but fragile politics. Some success by southern politicians in establishing relations with China and the USSR triggered a campaign of assassination and international terror by the northern government against southern leaders in 1983. South Korea's economic strength, American and Japanese support and the continued courting of the Soviet and Chinese governments are, however, likely to protect its separate existence against northern ambitions. In 1984 the diplomatic interaction between North and South was mostly positive.

For the USA a major cause for unease lies off the Asian shore in the unsteadiness of the Marcos regime. Opposition has risen among the middle classes of Manila, adding to the antipathy of Muslims and insurgents in the neglected southern islands of the group. The prospect of political breakdown and a radical change of rulers puts a major American foothold at risk. In some provinces the guerrillas of the New People's Army are firmly in command. Should the fortunes of local politics turn against the USA, the Subic naval base and Clark airforce base and, thus, the ability of US military power to reach into the Indian Ocean and counter the Soviet presence in Cam Ranh Bay in Vietnam, would be in jeopardy. More importantly, perhaps, the identification of the established order in the Philippines with the USA means that a good deal of face could be lost.

Disengagement

The main danger to our survival is the two-sided nuclear confrontation of the USA and USSR, competing for global hegemony. The threats and counterthreats of nuclear strategy arose originally in their contest for influence over Europe. Most recently there has been an increase in the force they confront each other with in Eastern Asia, including the deployment of nuclear weapons. The economic significance of Japan, South Korea, Hongkong and Taiwan and the potential of China is shifting the balance of American attention more strongly from the North Atlantic basin to the Pacific. Western Europe's global significance is declining comparatively. In July 1984 Secretary of State Schultz voiced this inclination when he said: 'The Pacific and the future are inseparable.'

If they were to stand back from this contest for the attention of the world, the leaders of both the USA and USSR would see that their real strength and security lie in their geography, not in nuclear armaments. These weapons, devised to deter what they each perceive to be the other's ambitions, have only negated their geographical security and raised the significance of every local conflict they get involved in to the level of a potential world-destroying spark. The hope of disengagement and of the surer survival of mankind must depend on an acceptance of the safety which distance affords against conquest and of the impossibility of achieving peace of mind by nuclear deterrence.

In thoroughgoing analyses of the nuclear strategies of the USA and USSR, Schell and Dyson have clarified the moral, technical and practical uncertainties which put our existence at risk. Dyson asks what the nuclear weapons are deemed to be for and reveals a potentially deadly incompatibility between American and Soviet strategic concepts.

The American establishment's reason for having these weapons is captured in the phrase 'assured destruction'. The ultimate objective is the destruction of the enemy. The threat made is that in the last resort, if the Soviets drive America to war, they will not survive. The main end of US military policy, then, is to assure the ability to destroy the USSR as a society no matter what kind of attacks they can launch on the USA. The assumption that the Soviets had the same basic strategy, and that the horrifying prospect of mutual assured destruction would hold an all-out exchange of nuclear fire at bay, was quite false. The Soviets have stuck by a counterforce strategy from the outset. The Russian tradition is one of defence and survival. The rulers of the USSR have stated time after time that they reject the notion of mutual assured destruction. They will not recognise the prospect of a limited nuclear war, nor do they countenance the use of nuclear weapons for the deliberate destruction of the civilian population. Looking from the Russian reaches of the European plain, a pre-emptive first strike at American missile silos in the face of imminent attack is reasonable, whereas any first use of tactical or theatre nuclear weapons is unacceptable since it endangers the lives of Soviet citizens and their allies immediately. Seen from the banks of the Potomac or near the beach in Santa Monica, first use of tactical weapons in Europe may keep the holocaust under control by bringing the Soviets to negotiate and will certainly not threaten American civilians. But from this vantage point a pre-emptive first strike hits at the very nub of American strategy and is anathema in public discussion. According to Ford, however, the first-strike option has been included in all US operational war plans since 1960.

Given this disparity of positions, there is no configuration of nuclear weaponry on both sides which will satisfy both strategic objectives simultaneously. Any efforts to limit armaments by agreement must founder on this underlying discordance. The question then is: given the nature of man, is there a strategic stance for one side which will defend self-interest without putting the lives of most of mankind at risk? Given an uncertain world in which we are ignorant of the future, this strategy must be robust. By robust, we mean that a strategy must demonstrably produce an acceptable outcome under every conceivable future state of the world. Future states of the world include a variety of strategies which the enemy may employ.

In addition to assured destruction and counterforce, there is a strategy which involves preparation to fight a limited nuclear war. Evidently the official statement of US nuclear strategy, 'Presidential Directive 59', approved by Carter in 1980 and accepted by Reagan, gives the prospect of fighting a limited nuclear war the same standing as assured destruction. There is thus an inbuilt source of confusion at the centre of the American war machine. The Russians refuse to accept that nuclear war could be limited to selective or tactical exchanges. They have stated that they do not accept the distinction between tactical and strategic weapons and will meet any use of nuclear weapons with a full-blown, counter-force riposte.

A technological fix strategy, spending lots of money on 'Star Wars' defences, has been given much publicity. The US areospace industry and airforce has been commissioned to produce mostly non-nuclear, defensive systems, extending the confrontation into space. Security would rest on the ability of US scientists and technicians constantly to outstrip their Soviet counterparts. This introduces a large element of uncertainty into the balance of power. The attention paid to defences does not abolish offensive weapons. The 'Star Wars' strategy remains a nuclear-war-fighting policy and thus remains prone to an accidental start to catastrophe.

A policy which might be made appealing to soldiers is that of giving up nuclear weapons unilaterally and depending on conventional forces, with enhanced non-nuclear technology, for defence. This would involve a return to traditional notions of honourable war, that did not depend on threatening to kill civilians to achieve its end. The main argument which can be mounted against this is that the knowledge of how to unleash nuclear fire cannot be buried and that the USSR would not necessarily follow suit if the USA disarmed. This allows the prospects of proliferation of nuclear weapons and of nuclear blackmail by the armed against the unarmed. If we accept that the desire to dominate others and hatred are deeply etched in people's souls, then we must ask what would stop the Soviets from wiping out the USA in cold blood? What can be set against this likelihood is the fact that they have always claimed to aim nuclear weapons or related targets. There would be no material gain to the Soviets from such an action. They would lose any moral authority they had acquired in the world. Soviet society might, indeed, suffer a considerable loss from the destruction of the USA, wrought by fallout, smoke and dust. If a strategy based on non-nuclear force gained enough of a following to support a bid for political power, it might well provide a basis for a steady balance of force in the world.

The extreme strategy of pacifism, of offering only non-violent resistance to aggression, stands little chance of political acceptance and must remain an ideal of personal and political behaviour only occasionally approached in practice. In modern industrial societies it faces the deeply entrenched vested interests of the military establishment, their industrial providers and all of their employees, dependants and servers. Pacifism would require all of these to seek employment elsewhere without the patronage of the sacred cow of national defence, which has extracted such lavish sacrifice in the past.

Having explored these feasible options, Dyson concluded by proposing a middle way between the nuclear and non-nuclear strategies which he called 'live and let live'. This involves the adoption of an explicit set of decision criteria by the USA in planning and acting for its defence. These were based on testimony given before a congressional committee on national security in 1969 by Donald Brennan, a strategic analyst. He was opposing the development of multiple-warhead weapons and the strategy of assured destruction. Brennan enunciated two principles of defence. The first was that the US military maintain the ability to damage the USSR as badly as the Soviets can damage the USA. The second required that, whenever the choice arose, live Americans be preferred over dead Russians. If the Soviets have a complementary preferrence for live Russians over dead Americans, then there is a mutual interest. The Soviets have talked and acted in a manner consistent with this criterion. The mutality would be that each side prefers its own protection to the enemy's destruction. This undercuts the validity of assured destruction as a permanent strategic position. It provides the basis for negotiating a reduction in nuclear weaponry down to the level where geographical strength becomes significant and where the world of nuclear power becomes a latently safer oligopoly, rather than the present upwardly unstable duopoly. With oligopolistic competition the logical argument for total nuclear disarmament becomes cogent. Two competitors will always be tempted to seek superiority over each other. If three or more nuclear powers in negotiation simultaneously require that they be allowed nuclear firepower equal to that of the combined strength of their enemies, then the only mathematically correct solution is for each to have no firepower.

If the rulers of the USA saw the value of adopting Brennan's principles for playing the power game, they would be in a position to realise more clearly the protection that their geography affords them. The USSR could then also relax in this security. Both of these nations have immense resources of space to absorb and waste the energies of would-be conquerors. Although the USA or the USSR could be destroyed by nuclear weapons, neither could be conquered. No opponent could muster the troops and resources to win and hold the territory of either nation. Russia has been attacked from the West in the era of mechanised warfare, but could not be overwhelmed. There is a difference between the two inasmuch as the USA has oceans isolating it from any conceivably effective invader. The USSR does not have the inviolable isolation of the USA. The strategists of the USA should, then, make an allowance in their calculations for the greater feeling of vulnerability that geography imposes on the Soviets. But neither power needs to be as nervous of their safety as their current strategic positions imply, if their objectives are solely defensive.

The idea of communists invading the USA from Mexico or Cuba, the ultimate bogeyman of domino theory, is laughable. Any would-be conquerors or liberators would be faced with an almost totally hostile population with an abundance of guns. A society in which people shoot each other in record numbers could really make a mess of uninvited strangers. By the same token a capitalist invasion of the USSR would be bound to fail. The lessons of the past, the expanse and climate of the USSR and the traditions of the Russian people and military are set against such an eventuality. Even if the conventional forces of the USA or USSR were defeated, guerrilla war, the offensive defence of one's home ground, is incredibly effective. To overwhelm partisan resistance requires a ratio of at least ten regulars to one insurgent, and much more than this in intensely farmed, wooded, mountainous or urban terrain. If the leaders of the USA and USSR could relax in confidence of their geographical indigestibility, then we might all be allowed to sleep easier, eventually.

In this ultimate test of our capacity for moral behaviour, which poses the problem of learning to live with each other in peace, distance does turn out to have a postive value. Although it is costly to overcome and has fostered the divergence of race and culture, which we use to excuse our lust for power, if we let it, distance can soften the clash of our ambitions. The tyranny of distance can provide us with freedom from each other, nurturing the delightful profusion of geographical variety, which makes life interesting and worth preserving. With nuclear weapons the saving grace of distance is denied the great powers. They can win it again if they find a way to put these arms aside.

Readings

The best and most recent appraisals of our nuclear predicament are:

Dyson, F. *Weapons and Hope* (Harper/Bessie, New York, 1984)

Ehrlich, P., C. Sagan, D. Kennedy and W. Roberts *The Cold and the Dark: The World after Nuclear War* (W.W. Norton, New York, 1985)

Ford, D. *The Button: The Pentagon's Strategic Command and Control System* (Simon & Schuster, New York, 1985)

Schell, J. *The Abolition* (Knopf, New York, 1984)

The events leading up to this state are examined in:

Barnet, R. 'Alliance', *The New Yorker*, 10 and 17 October 1983

The evolution of US nuclear strategy is chronicled in:

Kaplan, D. *The Wizards of Armageddon* (Simon & Schuster, New York, 1983)

The Strategic Arms Initiative is treated critically in:

Drell, S., P. Farley and D. Holloway *The Reagan Strategic Defense Initiative: A Technical, Political and Arms Control Assessment* (Stanford University, Stanford, 1984)

Jasani, B. (ed.) *Space Weapons — The Arms Control Dilemma* (Taylor & Francis, London, 1984)

Tirman, J. (ed.) *The Fallacy of Star Wars* (Vintage, New York, 1984)

A collection of positive reactions can be found in:

Ra'anan, U. and R. Pfaltzgraff (eds) *International Security Dimensions of Space* (Archon, Hamden, Conn., 1984)

For the current state of world affairs I relied on *Foreign Affairs*, *The Economist*, *Time* and other news media.

INDEX

Printed and bound by CPI Group (UK) Ltd, Croydon, CR0 4YY

22/10/2024

01777621-0020